LETTERS ON
WAVE MECHANICS

ALBERT EINSTEIN

Letters on Wave Mechanics

Correspondence with H.A. Lorentz,
Max Planck, and Erwin Schrödinger

Edited by
K. PRZIBRAM

NEW YORK

Cover design by Milan Bozic
Text design and composition by India Amos

978-1-4532-0468-9

Philosophical Library
119 W. 40th Street
New York, NY 10018
www.philosophicallibrary.com

Distributed by Open Road Integrated Media, Inc.
345 Hudson Street
New York, NY 10014
www.openroadmedia.com

Contents

LETTERS 19–21

Lorentz & Schrödinger

Foreword

A GREAT PHYSICAL THEORY like Schrödinger's wave mechanics, when it is confirmed, takes on its own impersonal existence in the course of time, becomes completely detached from its originator, and is finally received as self-evident. In this way one forgets how many inner struggles, hopes, and disappointments were bound up with its beginnings and one forgets too all the pros and cons of contemporary reactions to it. This more personal side can be reawakened into life if there are contemporary letters like the ones reproduced here.

Schrödinger's widow, Mrs. Annemarie Schrödinger, cherished the wish that her husband's correspondence concerning wave mechanics might be published among the works of the Austrian Academy of Sciences and so be made accessible to a wider scientific circle. She turned to the undersigned, as the senior among Austrian physicists, with the request that he make her wish known to the Academy. A motion concerning the publication of the letters was passed unanimously and with joyful gratitude at the meeting of the Academy's Division of Mathematical and Natural Sciences on 25 January 1962; the editing was entrusted to the undersigned.

Little needs to be added to the letters; they speak for themselves. Apart from their essential content, they reveal something of the personalities of the four men of genius, corresponding to Buffon's sentence, "Le style c'est l'homme."

There are some omissions in the carbon copies of

Schrödinger's letters, which were all that were available here, since the mathematical formulas that were entered by hand on the typed originals are often missing. These omissions were supplied according to the meaning and by comparison with Schrödinger's published works. The other scientists' communications are all in the form of hand written letters, or postcards (No. 1, 7, 10, and 12). A calculation on wave packets that filled many pages was omitted from Lorentz's second letter (No. 21); also omitted were the beginnings of letters 8, 15 and 16, which contained only personal matters, and a paragraph in letter 8 dealing with molecular statistics. The sketch in No. 12 is a facsimile in natural size. All texts are reproduced faithfully—*salve errore et omissione*; several inconsistencies in punctuation and style have been left uncorrected. Some (numbered) footnotes, set in smaller print, may be of assistance in giving a broader orientation.

We express our thanks to the heirs of Max Planck and H. A. Lorentz as well as to the Executor of the Estate of Albert Einstein for permission to publish the corresponding letters, and to the latter also for photographic copies of letters 13 and 15, (as well as the letter referred to in the footnote to letter 13), no carbon copies of which were to be found here.

Finally we thank the Springer-Verlag of Vienna for undertaking the publication and for its painstaking accomplishment.

K. PRZIBRAM
Vienna, Summer 1963

Introduction

___

"IN THIS ARTICLE I should like to show, first of all for the simplest case of the (non-relativistic and unperturbed) hydrogen atom, that the usual rule for quantization can be replaced by another requirement in which there is no longer any mention of 'integers.' The integral property follows, rather, in the same natural way that, say, the *number of nodes* of a vibrating string must be an integer. The new interpretation can be generalized and, I believe, strikes very deeply into the true nature of the quantization rules." With these words Erwin Schrödinger began the first paper of his series, "Quantization as a Proper Value Problem," sent off to the *Annalen der Physik* at the end of January, 1926.[1] By the end of June he had completed four more major papers developing and applying the concepts and methods of a new wave mechanics that he hoped would be related to classical mechanics in the same way that wave optics is related to geometrical optics. What impressed him most in his elegant theory, perhaps even more than its evident power to treat a wide range of basic atomic problems, was its "naturalness," its apparently intuitive character for anyone at home in classical physics, and the way in which it seemed to avoid the most perplexing and disturbing features of the existing quantum theory.

For Schrödinger was writing a quarter of a century after Max Planck had broken with the past by introducing energy quanta into physics, in order to explain the black-body radiation law. During those twenty-five years physicists had been confronted with a series of shocking departures from established modes of thought: Planck's treatment of the energy as a discrete rather than a continuous variable was followed by Einstein's modest proposals that radiation must be viewed as somehow composed of independent particles of energy and that a quantum theory of matter as well as radiation must be constructed. In 1913 Niels Bohr compounded these heresies in a theory that explicitly denied the validity of electrodynamics for atomic radiation processes and made the frequencies of atomic spectral lines independent of the frequencies of electronic motions within the atom. Bohr's ideas, strange as they seemed, served as the starting point for a serious and partly successful attempt to construct a theoretical structure that could explain the physical and chemical properties of matter, including the mysterious regularities recorded by the spectroscopists. By the spring of 1925 the theoretical picture had been elaborated by the work of many physicists into a tantalizingly incomplete and confused tangle of successes and failures, so that Wolfgang Pauli, one of the most acute, and most outspoken, of the younger theorists could write to a friend: "Physics is very muddled again at the moment; it is much too hard for me anyway, and I wish I were a movie comedian or something like that and had never heard anything about physics!"[2]

Within a few months the atmosphere changed abruptly. Werner Heisenberg, in Göttingen, proposed a new approach to the riddles of the quantum theory and this new approach was quickly developed into an elaborate mathematical formalism by Max Born, Pascual Jordan and Heisenberg himself.[3] The new theory—called quantum mechanics by its authors but often referred to as matrix mechanics after its principal mathematical technique—gave promise of really providing the beginnings of a *consistent* quantum theory, for the first time. It did this, however, only at the price of an even sharper and deeper break with the past, giving up any attempt to offer a physical or intuitive picture of the processes whose outcome could be calculated, and requiring that the theory deal only with relations among quantities that could, at least in principle, be observed.

This brief and necessarily oversimplified account may give some impression of the situation in theoretical physics when Schrödinger's work began to appear in the spring of 1926. I have deliberately emphasized the widespread sense of the strangeness and even the arbitrariness of the quantum theory, because it was just these properties of the theory that Schrödinger was so happy to avoid with his new wave mechanics. This happiness is evident in his papers, and especially in the paper, *On the Relationship of the Heisenberg-Born-Jordan Quantum Mechanics to Mine*,[4] in which to his own surprise and joy Schrödinger was able to demonstrate the complete mathematical equivalence of these two theories, so different in their starting point, their method, and their spirit.

Most of the letters in this volume date from the spring of 1926 and give the reactions of Planck, Einstein, and Lorentz to Schrödinger's ideas. Planck was obviously enraptured by Schrödinger's work. Always a traditionalist, essentially conservative in his views, despite his having conceived the revolutionary idea of the quantum of energy in 1900, he welcomed just that "natural" or "intuitive" aspect of wave mechanics which so appealed to its creator.

Lorentz's reaction to Schrödinger's work is especially remarkable. For many years physicists had always been eager "to hear what Lorentz will say about it" when a new theory was advanced, and, even at seventy-two, he did not disappoint them. His letter of May 27, 1926 consists of a long analysis and critique of Schrödinger's work in which Lorentz puts his finger on some of the most questionable points: the dispersion of wave packets, the difficulty of interpreting Schrödinger's waves in a system of more than one particle, and the doubtful aspects of Schrödinger's proposal that radiation be understood as a kind of beat frequency phenomenon. Lorentz's letter, Schrödinger's detailed response, and Lorentz's rejoinder (see letters 19–21) give a clear picture of the difficulties that worried Schrödinger as he tried to develop his theory further.

It should be pointed out that not all theorists shared the enthusiastic views of Schrödinger and Planck that wave mechanics at last indicated the proper direction for future theory, or even the more temperate opinion of Lorentz that it would be a pity if this did not turn out to be the right direction. Heisenberg was "deeply disturbed" at the attempt to make the wave concepts central in the theory and expressed himself in no uncertain terms in a letter to Pauli: "The more I reflect on the physical part of the Schrödinger theory the more detestable I find it. Schrödinger really simply throws overboard everything in quantum theory: namely, the photoelectric effect, the Franck [-Hertz] collisions, the Stern-Gerlach effect, etc. Then it isn't hard to make a theory."[5] And, when Schrödinger lectured on his work at Copenhagen in September 1926, Bohr tried very hard to persuade him that the discontinuous transitions were really indispensable; to which Schrödinger responded: "If we are going to stick to this damned quantum-jumping, then I regret that I ever had anything to do with quantum theory."[6]

It was the fate of Schrödinger's ideas to be absorbed into

the new synthesis of the following year, the Copenhagen Interpretation of quantum mechanics, developed principally by Heisenberg and Bohr, and based on Born's statistical interpretation of the wave function. Schrödinger never liked the Copenhagen Interpretation and, especially in his later years, marshalled his keen insight, vast erudition, and formidable literary ability in a series of sharp attacks on the prevalent views.[7] The few letters in this volume that date from the period after 1926 suggest some of his feelings and ideas on this subject. It is no accident that they form a part of his correspondence with Albert Einstein.

Einstein's first response to Schrödinger's work (see letter 9) is as characteristic and, in its own way, as remarkable as that of Lorentz. Einstein had apparently misread Schrödinger's first article, or misremembered it when he thought about it later, and was dissatisfied with the equation he thought to be Schrödinger's. This equation failed to satisfy two critical but very general properties: the allowed energy values of independent systems ought to be additive and the equation ought not contain the arbitrary integration constant in the energy. Since the equation Einstein thought to be Schrödinger's did not have these properties, Einstein suggested another that did: it was just the equation that Schrödinger had in fact introduced in his own paper. Schrödinger was delighted and took Einstein's remarks as providing new evidence for the reasonableness of his method.

Schrödinger's wave mechanics owed much to Einstein's earlier work. One usually reads that Schrödinger was inspired by Louis de Broglie's thesis in which the concept of matter waves was first advanced, but that is hardly an adequate account. It was Einstein's profound studies of the wave-particle duality for radiation that originally suggested to de

Broglie that a corresponding duality should exist for matter. And it is not surprising that it was Einstein who first recognized the importance of de Broglie's brilliant idea. "Read it," he said to Max Born of de Broglie's thesis, "even though it might look crazy, it is absolutely solid."[8] He was also the first to see the implications of de Broglie's matter waves and to offer new arguments for their existence in his own quantum theory of the ideal gas (the Bose-Einstein gas).[9] Schrödinger's study of this important development in statistical mechanics drew his attention to de Broglie's work, as he himself pointed out several times in his papers and in his letter to Einstein of April 23, 1926, reproduced here. Without Einstein's "short but infinitely far-seeing remarks" Schrödinger might never have tried to develop de Broglie's ideas further into a full fledged wave mechanics.

Neither Schrödinger nor Einstein had ever taken a major part in the development of the "old quantum theory" with its strong emphasis on applying quantum rules to problems of atomic structure and atomic spectra. This fact may not be irrelevant to the distaste that both men later felt and expressed for the Copenhagen Interpretation. Einstein's often stated opinion that the quantum mechanical description of physical reality could not be considered complete was reinforced by Schrödinger's clever conceptual experiment involving the "quantum mechanical cat." (See letters 16-18) Both men felt, rightly or wrongly, that the great majority of their colleagues had chosen the wrong path.

Schrödinger expressed his concern that physics was running "the grave danger of getting severed from its historical background."[10] During the years he spent in Dublin he often lectured and wrote on this theme. I think he would have liked this little collection of his correspondence, this fragment of a critical chapter in the history of science, since he himself liked to quote, with evident approval, these words of Benjamin Farrington: "History is the most fundamental science, for there is no human knowledge which cannot lose its scientific character when men forget the conditions under which it originated, the questions which it answered, and the functions it was created to serve."[10]

MARTIN J. KLEIN
Case Institute of Technology
Cleveland, Ohio

Schrödinger & Planck

1

Planck to Schrödinger

Berlin—Grunewald

2 April 1926

Dear Colleague,

Many thanks for the reprint. I read your article[1] the way an inquisitive child listens in suspense to the solution of a puzzle that he has been bothered about for a long time, and I am delighted with the beauties that are evident to the eye, but I have to study it much more closely and in detail to be able to grasp it completely. Besides, I find it extremely congenial that such a prominent role is played by the action function W. I have always been convinced that its significance in physics was still far from exhausted. There is just one little blemish that I would have been glad to see removed. Old Jacobi would have been a little annoyed, despite all his interest, over the alteration of his name.[2] Can it still be changed?

Yours,
Planck

2

Schrödinger to Planck

Zürich

8 April 1926

My dear Professor,

I was indescribably delighted by your kind card of April 2nd. I am especially happy that the basic idea seems plausible to you, and am now very confident that in the course of time it will be worked out in a way that is useable in all respects, no matter how imperfect it may be at present.

I am very ashamed about the dreadful "k," and immediately wrote to the printers; I hope it can still be changed. Many thanks—the worst of it is the ironclad consistency with which I disfigured this hallowed name in *five* places; it would have been terribly distressing to me.

Thank you very much for kindly sending me your lecture,[3] which I had already read with the greatest interest several days earlier. I was especially captivated by the dramatic force with which you sketch the status of the theory of relativity and the quantum theory—in the third section—and with the way you pick out the key difficulty and make it comprehensible without formulas. Just this difficulty concerning the energy unfortunately still persists, quite unimpaired.

If I did not answer your card, which gave me so much pleasure, at once, it was because I wanted to send along at least a little something that was new. Enclosed are the results for the Stark Effect in H. It seems that the intensities come out completely right. The assumption on which it is based is that the electrical charge density is given by the square of the wave function, and that the normalization integral has the same value for all the *individual* proper vibrations that belong to *one* coarse Balmer level. I cannot yet describe the numbers I am sending you as incontestable because the calculation is very involved and I have not yet checked everything again. In any case Epstein's formula for the splitting comes out completely unaltered (as I already said at the end of my "Second Paper"); also the "Selection Rule for the azimuthal quantum number." Moreover, the "exclusion of zero for the equatorial quantum number" also comes out quite automatically—*there is no* proper vibration that would correspond to the quantum orbit that collides with the nucleus. It is also very gratifying that although the three unobserved components at relative distances of 5, 6, and 8, are not actually "forbidden" theoretically, they receive an intensity that is 80 to 700 times smaller than that of the weakest observed component, so that their non-appearance becomes very understandable.

I am now calculating H_α, H_β, H_γ. The calculations are unfortunately terribly difficult to see through and I cannot manage to bring them into a simpler form.

With best compliments and greetings I remain,
dear Professor, always
Yours faithfully,
E. Schrödinger

3

Planck to Schrödinger

Berlin—Grunewald

24 May 1926

Dear Colleague,

I have owed you my thanks for sometime for your kindly having sent me your last *Annalen* article on quantization. You can imagine the interest and enthusiasm with which I plunge into the study of these epoch making works, although I now make very slow headway penetrating into this peculiar train of thought. In connection with that I have high hopes of the beneficial influence of a certain amount of familiarity which in time facilitates the use of new concepts and ideas, as I have often found already. But what especially delights me, and the reason for my really writing you today, is the joyful hope that we may soon have the opportunity to hear you and to talk to you here. As my colleague Grüneisen[4] tells me, your visit to a meeting of the Physical Society has not been cancelled but only somewhat postponed, and it may even still take place this semester. Let me tell you explicitly how much pleasure all of the physicists here would have in hearing you yourself present your new theory and in coming into contact with your ideas. And don't be afraid that we will make too many demands on you and tire you out. I do not know if you are already familiar with Berlin. But I hope you will find that in certain respects life here is freer and more independent than in a smaller city where everyone checks on everyone else, and there is no possibility of completely withdrawing at some time without anybody noticing it.

I should like to express just one little selfish request. In case you can come in July, please not before the 11th. Because at the beginning of July I have to go to Bonn for a few lectures and I would be sad if I missed your visit here as a result. Above all, however, I wish you the relaxation that you need after your demanding labors, and the complete recovery of your powers. I should be especially grateful if, at your convenience, you would send me a brief card with a word about your travel plans.

In the meantime, with best regards,
Yours sincerely,
M. Planck

4

Schrödinger to Planck

Zürich
31 May 1926

My dear Professor,

Thank you very much for your kind and extremely gracious letter of the 24th, which now has finally decided me to accept the attractive invitation for this semester, however things may go. I have just written to Mr. Grüneisen. It goes without saying that, so far as I am concerned, a date when you are absent from Berlin is out of the question. Now Mr. Grüneisen was kind enough to point out to me that it might also be possible to consider a slight postponement of the date of the meeting, and since a postponement of the July 9th meeting would surely come too near the end of the semester, as he himself thinks, I have allowed myself to suggest that perhaps the June 25th meeting could be put off until July 2nd. Would that still work out with your trip to Bonn? The 25th of June would not be acceptable to me because from the 21st to the 26th a number of foreign physicists (among them Sommerfeld, Langevin, Pauli, Stern, P. Weiss) are meeting here for lectures and discussions. Now the connections work out so badly that I would have to leave here on the afternoon of the 23rd at the latest, if I do not want to travel through the night directly before the Berlin meeting. And I should not like to do that because then I am often completely exhausted and may possibly speak *very* badly.

I should be very grateful if you would give me some hints, in just a few words, as to how I should plan my lecture. What I mean is, should I think more of the fact that you and Einstein and Laue are in the audience—a thought without which I should feel uneasy—or should I direct myself more to those gentlemen who are further removed from theoretical work; which would of course have as an inevitable result that those named above (and a considerable number of others) will be very bored. In other words: should I recapitulate in a simplified way what has already been published or, passing over that lightly, talk more about perturbation theory, the Stark effect, and general intensity formulas? (Otherwise I could only mention these latter things briefly at the end, or else it would get to be too long; it takes about an hour for a general survey of the fundamentals, for the purpose of orientation and without much calculation, as I know from our colloquium here).

Naturally I can also do both, if there is the opportunity, one in a general meeting and the other in a more restricted colloquium.

Today I received a very kind and very interesting letter of 13 closely written pages from H. A. Lorentz[5] which I still have to study in detail, of course. He raises a good many interesting questions; however, he does not reject it at all, on the whole, but still appears to be very critical. Lorentz sees one of the chief difficulties in reinterpreting classical mechanics as "wave mechanics" to lie in the fact that the "wave packet" which is to replace the "representative point" of classical mechanics in macroscopic problems, (possibly also in the motion of the electron on paths of slight curvature), that, I say, this wave packet will *not remain together*, but, on the contrary, will gradually spread into larger volumes by "diffraction," according to general theorems of wave theory. I felt that to be a serious point at first—yet, strange to say, it seems *not to be the case*, at least not always. For the harmonic oscillator (which always remains the simplest typical example of a mechanical system which one can work with so easily and agreeably), I was able to produce a wave packet, by superposition of a large number of neighboring characteristic oscillations of high order (i.e. high quantum number), which is practically confined to a small spatial region, and which as a matter of fact revolves in precisely the harmonic ellipses described by classical mechanics for an arbitrarily long time *without* dispersing! I believe that it is only a question of computational skill to accomplish the same thing for the electron in the hydrogen atom. The transition from microscopic characteristic oscillations to the macroscopic "orbits" of classical mechanics will then be clearly visible, and valuable conclusions can be drawn about the phase relations of adjacent oscillations. For the present these phase relations and amplitude relations remain postulates, however; they can naturally also be so arranged that for large quantum numbers a "revolving" mass point does *not* result: e.g. since the structure is linear it can also be arranged so that *two* wave groups, revolving independently of one another, result—perhaps the equations are only approximately linear.

A second very delicate question that concerns Lorentz is the energy that is to be assigned to a characteristic oscillation. It is quite certain that the Balmer-Bohr energy value is not to be ascribed *to the characteristic oscillation*. In general one should not consider the individual characteristic oscillation as the equivalent of the individual Bohr orbit; that is a mistaken parallel, as the above construction shows. The concept "energy" is something that we have derived from macroscopic experience and really *only* from macroscopic experience. I do not believe that it can be taken over into micro-mechanics just like that, so that one may speak of the energy of a single partial oscillation. The energetic property of the individual partial oscillation is *its frequency*. Its *amplitude* must be determined in quite another way—I believe by normalizing the integral of the square of the total excitation to the value of the electronic charge.

Mr. Grüneisen was kind enough to hold out to me the prospect that either you or Mr. von Laue would offer me hospitality. If it doesn't cause too much trouble I am naturally very pleased about it, and in any case I am very grateful for your kind offer. I would strive to give as little inconvenience as possible, and ask that it be so arranged that you are disturbed as little as possible; naturally any improvised lodging you choose is completely adequate for me.

Thank you once again for all the kindness that is always shown me by Berlin in general and by you especially, Professor Planck.

With sincere respect, I remain
Yours faithfully,
E. Schrödinger

5

Planck to Schrödinger

Berlin—Grunewald

4 June 1926

Dear Colleague,

I am extremely pleased that you could make up your mind to visit Berlin before the end of this semester, and I know for certain that the rest of the physicists here think the same way.

My colleague Grüneisen informs me that he has some doubts with regard to July 2nd and suggests July 16th instead. I should just like to join him in this. The semester here lasts until the beginning of August so that things are still in full swing in the middle of July and we need not be afraid that many will have already gone away. Grüneisen himself is an exception, to be sure, but he has to set out so early that he would unfortunately miss your visit all the same. But the 16th of July would suit the rest of us very well, and the only question is whether it is suitable for you yourself.

My wife and I would be especially happy if you would stay with us. We hope very much that we will be able to make you comfortable in our house. I shall take care above all that you remain master of your own actions to the greatest possible extent, and especially that, at those times over and above the "official" periods dedicated to the Physical Society, you have the opportunity to withdraw and to occupy yourself as you see fit. I know from experience how pleasant it often is to have a possibility of this kind. Moreover, my house stands at your disposal night and day for as long as you are inclined to stay.

You also talk about the level at which your lecture should best be given, or rather at which it should begin. I would like to propose, in agreement with my colleagues, that you imagine your audience to be students in the upper classes who, therefore, have already had mechanics and geometrical optics, but who have not yet advanced into the higher realms; to whom, therefore, the Hamilton-Jacobi differential equation, *if* they are acquainted with it at all, signifies a difficult result of profound research, deserving of reverence, and not by any means something to be taken for granted. Under no circumstances, however, should you be afraid that any one of us will consider one sentence of yours to be superfluous. For even if the sentence should not be necessary for an understanding of your train of thought, it would always offer the particular interest of seeing what special paths your thought takes and which particular forms your perception favors. For all of us the main point of your lecture will be what you yourself in your letter designated as a general survey of the fundamentals for the purpose of orientation without much calculation and without many individual problems. Perhaps it would be easier and more natural for you to carry this out, if on the other day, Saturday morning the 17th of July, you were to give a second lecture in our Colloquium, aimed at more special matters with supplements and continuations of the lines of thought you will have described at the more general meeting. I hope that this seems suitable to you, since you already indicated such a possibility yourself. That can very easily be arranged, and I ask you only to let me know so that we can take care of matters.

What a cross-fire of critical, enthusiastic, and questioning acclamations might now besiege you! But still, it is a thing with incredible prospects. I see that you have already energetically taken hold of the big question of whether and under what conditions a wave packet will remain intact. I have such a feeling that for closed systems it is the boundary conditions that take care of the conservation [of the wave packet], whereas a satisfactory solution for phenomena in an unbounded space seems to me to be possible only on the basis of new assumptions. That, however, is a *cura posterior*.

In the meantime my cordial greetings and the friendly request that you write me the day and hour that you arrive here.

Yours faithfully,
Planck

6

Schrödinger to Planck

Zürich

11 June 1926

My dear Professor,

Please do not be annoyed with me because I am just today answering your extremely kind letter of the 4th of June. I have written to Mr. Grüneisen in the meantime that I now finally accept for July 16th, and in fact it also suits me excellently because then I need to conclude my lectures a few days earlier, and besides, these last lectures are no longer worth much, since the men already have their heads full of the vacation. I am very sorry, however, not to be able to see or to become acquainted with Mr. Grüneisen himself, but unfortunately that can't be helped.

Now first and foremost my very hearty thanks for your kind invitation to stay with you which I of course accept with the utmost pleasure. The words with which you offer me your house as a "place of refuge from Berlin" express a boundless, thoughtful, concerned kindness that has truly touched me. You are quite correct that one is most often in want of just this possibility of being alone for a few hours in situations where everyone around is striving to be nice to one. I hope, however, that I will not need to make much use of this possibility in the present situation, despite my end-of-semester fatigue. Not only would I really like to give as much as I possibly can, both in and outside the "official" hours, to the gentlemen in Berlin who are so friendly as to be interested in my work; but also from a purely selfish standpoint I should like to make full and intensive use of the opportunity to discuss the things that have held me completely captured for months, with a number of the most distinguished scientists with the widest variety of research interests. If one still gets a little tired after a few days—the pleasure of the interesting dialogues would be sufficient compensation, to say nothing of the stimulation and the positive challenge.

I will hold to your advice for which I am very grateful, concerning the general lecture, and will naturally be very happy if anyone still has the desire to listen to me on the following day in the more restricted group.

By the way, during the last few days another heavy stone has been rolled away from my heart: I have the interaction of the atom with an incident light wave, thus the theory of dispersion. I had considerable anxiety over it because it was to be feared that the eigenfrequencies themselves would appear as the locations of the resonances in the case of a forced oscillation, and furthermore, that the forced vibrations would not depend on the existing nearby proper oscillations, i.e. not on the *state* in which the atom happens to be. And that would be nonsense. But it all resolved itself with unheard of simplicity and unheard of beauty; it all came out exactly as one would have it, quite straightforwardly, quite by itself and without forcing. This is the way: what I called the "wave equation" up to now is really not the wave equation but rather the equation for the *amplitude*. It no longer contains the time at all, but instead of it, it already has an integration constant E, (see Eq. [18″] of my second paper.) The time dependence must be given by $\psi \sim P \cdot R\left(e \pm \dfrac{2\pi iEt}{h}\right)$, or, what is the same thing, we must have

$$\frac{\partial^2 \psi}{\partial t^2} = -\frac{4\pi^2 E^2}{b^2}\,\psi$$

One can eliminate E from *this* equation and equation (18″) and one *thus* obtains the true wave equation which is of fourth order in the coordinates, perhaps of the type of the vibrating *plate*.

The main point is now this: one may now in a free and easy way also let the potential energy be an explicit function of the time in this true wave equation. The interaction energy with the incident wave can be added on as a perturbing term, and perturbation theory straightforwardly applied, which is quite simple. The result is essentially the so-called Kramers dispersion formula, with completely exact assertions about the phase and polarization of the secondary radiation, naturally assuming that the eigenfunctions and eigenvalues of the unperturbed atom are known.

What is still missing from the whole picture is only the interaction with its own wave, i.e. what corresponds to radiation damping. I believe that can no longer be very hard.

Naturally, perturbation theory can still be applied to many other questions too, e.g. the perturbation due to an α-particle or an electron flying past. I believe that it is a rather considerable step forward because the whole course of an event in time can now be exactly followed—at least in principle.

I should like to arrive in Berlin on the evening of July 15th, if that is agreeable to you; i.e. if it can be so arranged that the train does not arrive much too late. I will have to study the very many different possibilities first, and then I will let you know definitely. In the meantime, my warmest thanks once again to you and your wife for your great kindness. Please do *not* put yourself out at all; the less trouble I give you the happier I will be!

With sincere respect, I am always
Yours,
E. Schrödinger

7

Planck to Schrödinger

Grunewald
15 June 1926

Dear Colleague,

Many thanks for your letter of the 11th which brought us your most welcome acceptance. Besides that it once again communicates news that will make the heart of every theoretical physicist leap for joy. But we will be able to talk more about that when you are here; there are always many questions to be asked, for the appetite increases with eating. Just take care not to overwork. And so I await the announcement of your arrival time on the 15th of July. The earlier, the better. You lecture to the Physical Society[6] then on the 16th, and on the 17th to our Colloquium. On the evening of the 17th I hope to have several colleagues and you at our home. For all eventualities I repeat that I shall be in Bonn from the 5th to the 11th of July. My address, however, remains the usual one.

Warmest greetings.
Yours,
Planck

8

Schrödinger to Planck

Zürich

4 July 1927

My dear Professor,

. . . . . .

May I talk a little physics yet? I should like so much to know how the quantum situation is judged in Berlin and especially by you yourself. Is what the matrix-physicists and q-number-physicists say true—that the wave equation describes only the behavior of a statistical ensemble, just like the so-called Fokker partial differential equation perhaps? I would willingly believe it since the interpretation is really much more convenient, if I could only pacify my conscience and convince it that it is not frivolous to get off so easily in overcoming the difficulties. I believe I am right that you yourself wrestled with the first and most basic assumption of discontinuity (i.e. precisely "the quantum theory") in its day, wrestled a hard intellectual struggle with your whole soul, as the "second version"[7] which followed so long afterwards shows most clearly. I believe that one is obliged to take up this struggle anew with the same seriousness among today's newly emerged points of view. I do not have the feeling that this is really happening on the part of those who today already announce categorically: the discontinuous exchange of energy *must* be adhered to.

What seems most questionable to me in Born's probability interpretation is that when it is carried out in more detail (by its adherents) the most remarkable things come forth naturally: the probabilities of events that a naive interpretation would consider to be independent do not simply multiply when combined, but instead "the probability amplitudes interfere" in a completely mysterious way (namely, just like my wave amplitudes, of course). In a brand new article by Heisenberg even my much smiled at wave packets are said to have finally found their suitable interpretation as "probability packets." The first is especially comical. It can also be expressed this way: the Born probability (more correctly its square root) is a two dimensional *vector*; its addition is to be carried out vectorially. The multiplication is still more complicated, I believe.

Well, as God wills; I keep quiet. That is, if one really *must*, I too will become accustomed to such things.

With kindest regards to your wife, Professor Planck,

I remain

Yours faithfully,

E. Schrödinger

LETTERS 9–18
Einstein & Schrödinger

9

Einstein to Schrödinger

---

16 April 1926

Dear Colleague,

Professor Planck pointed your theory out to me with well justified enthusiasm, and then I studied it too, with the greatest interest. In the process one doubt has arisen which I hope you can dispel for me. If I have two systems that are not coupled to each other at all, and if E_1 is an allowed energy value of the first system and E_2 an allowed energy value of the second, then $E_1 + E_2 = E$ must be an allowed energy value of the total system consisting of both of them. I do not, however, understand how your equation

$$\text{div grad } \phi + \frac{E^2}{b^2(E - \Phi)}\phi = 0$$

is to express this property.

So that you can see what I mean, I put down another equation that would satisfy this condition:

$$\text{div grad } \phi + \frac{E - \Phi}{b^2}\phi = 0$$

For, the two equations

$$\text{div grad } \phi_1 + \frac{E_1 - \Phi_1}{b^2}\phi_1 = 0$$
(valid for the phase space of the first system)

$$\text{div grad } \phi_2 + \frac{E_2 - \Phi_2}{b^2} \phi_2 = 0$$
(valid for the phase space of the second system)

have as a consequence

$$\text{div grad } (\phi_1\ \phi_2) + \frac{(E_1 + E_2) - (\Phi_1 + \Phi_2)}{b^2}(\phi_1\ \phi_2) = 0$$
(valid in the combined q- space).

As proof one need only multiply the equations by ϕ_2 and ϕ_1 respectively and add. $\phi_1\ \phi_2$ would, therefore, be a solution of the equation for the combined system, belonging to the energy value $E_1 + E_2$.

I have tried in vain to establish a relationship of this sort for your equation.

It also seems to me that the equation ought to have such a structure that the integration constant of the energy does not appear in it; this also holds for the equation I have constructed, but despite that I have not been able to assign a physical significance to it, a matter on which I have not reflected sufficiently.

With warmest greetings from
A. Einstein

The idea of your article shows real genius.[8]

10

Einstein to Schrödinger

---

22 April 1926

Dear Colleague,

I have just seen from your first article that you really based your considerations on the equation

$$\text{div grad } \psi + \text{constant } (E - \Phi)\, \psi = 0$$

which satisfies the addition theorem for independent systems. So my letter was superfluous.

I see no basic difference between your work[9] on the theory of the [ideal] gas and my own. For according to you, too, the state (of equal probability) is characterized by the values of the set of numbers $n_1, n_2, n_3, \ldots$, where the numbers n_1, n_2, etc. have the same meaning as they do for me.

I do not understand how you are allowed to use the last form of (4), in your article, since this is not consistent with the condition $\Sigma n_4 = \text{const.}$

Best wishes from
A. Einstein

11

Schrödinger to Einstein

Zürich

23 April 1926

My dear Professor,

My hearty thanks for your extremely kind letter of the 16th. Your approval and Planck's mean more to me than that of half the world. Besides, the whole thing would certainly not have originated yet, and perhaps never would have, (I mean, not from me), if I had not had the importance of de Broglie's ideas really brought home to me by your second paper on gas degeneracy.[10]

The objection in your last letter makes me even happier. It is based on an error in memory. The equation

$$\text{div grad } \psi + \frac{E^2}{b^2(E - \Phi)}\psi = 0$$

is *not* mine, as a matter of fact, but my equation really runs *exactly* like the one that you constructed free hand from the two requirements of the "additivity" of the quantum levels and the non-appearance of the absolute value of the energy:

$$\text{div grad } \psi + 8\pi^2 \frac{E - \Phi}{b^2} \psi = 0$$

Your very basic requirements *are* therefore fulfilled. I am, moreover, very grateful for this error in memory because it was through your remark that I first became consciously aware of an important property of the formal apparatus. Besides, one's confidence in a formulation always increases if one—and especially if *you*—construct the same thing afresh from a few fundamental requirements.

Just recently I read with the greatest interest your proposal in *Naturwissenschaften* for a new coherence experiment.[11] I have not yet finished thinking it over. That always takes me rather long. I am not completely sure how you conceive of the arrangement behind the grating. ("Behind the grating the light will be made parallel by means of another lens . . .") I imagine the wire grating at the focus of this farther lens and then perhaps a Fabry-Perot interferometer (plane parallel layer of air, rings of equal inclination). Then one would usually say: each point of the light source corresponds to a point of the circular image. The grating is the light source. Then the light beams from different slits of the grating would really not interfere with one another. But, according to the classical theory, we have here the unique situation that different points of the light source vibrate coherently in a legitimate way. I have not yet made clear to myself how that works out. But maybe the arrangement, as I continue to think about it, is also stupid.

I very much enjoyed your delightful explanation of the formation of meanders.[12] It just happens that my wife had asked me about the "teacup phenomenon"[13] a few days earlier, but I did not know a rational explanation. She says that she will never stir her tea again without thinking of you.

Kindest regards from
Yours faithfully,
E. Schrödinger

12

Einstein to Schrödinger

---

26 April 1926

Dear Colleague,

Many thanks for your letter. I am convinced that you have made a decisive advance with your formulation of the quantum condition, just as I am equally convinced that the Heisenberg-Born route is off the track. The same condition of system additivity is *not* satisfied in their method.

I have now discovered considerations that nearly rule out the existence of elementary spherical waves, so that I am pretty well convinced that the experiment I proposed will turn out negatively. Its simplest realization is, in principle:

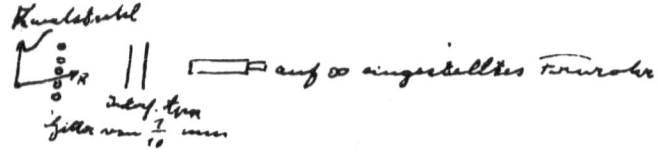

A direction of emission R corresponds to a point in the focal plane of the telescope. Rays emitted in the direction R by a particle reach, or do not reach, the telescope (alternately); for a suitable relationship between the particle velocity and the path difference the interference would have to be destroyed, which, however, I do not believe. Diffraction at the grating acts as a disturbance, but not so strongly as to destroy the demonstrative power of the experiment.

Friendly greetings.

A. Einstein

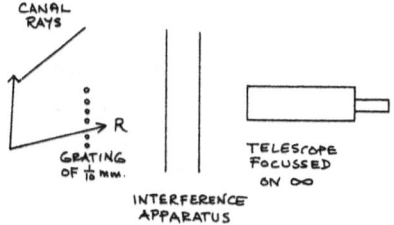

13

Schrödinger to Einstein

Cunostrasse 44
Berlin—Grunewald
30 May 1928

My dear Professor Einstein,

Enclosed is a letter from Niels Bohr[14] who, at the end, expresses the wish that you and Planck might also be made aware of its contents. I also enclose the carbon copy of my letter just so that you can see what set off the discussion. The remark about the uncertainty relation in the ideal gas runs as follows, when worked out: if we quantize a molecule that is reflected back and forth on the segment l, then we have

$$\oint p dx = p \oint dx = 2lp = nh; \quad \text{i.e.} \quad p_n = nh/2l.$$

Neighboring quantized values of the momentum therefore differ from each other by so little, (namely by only $h \backslash 2l$), that even with the largest possible uncertainty in the *coordinate* ($\Delta x = l$), I cannot buy enough accuracy in the *momentum* to allow me to distinguish between neighboring quantum states. What Bohr says about this case at the end of the third page, I do not understand at all.*

If it is agreeable to you, I would be glad to come over

---

* Everything that I have just said is surely terribly trivial!

sometime to talk about the letter, but perhaps you do not have much time now, before your departure, and you need to be sparing with it.

> With greetings and best regards to the whole family,
> Yours sincerely,
> *Schrödinger*

14

Einstein to Schrödinger

---

31 May 1928

Dear Schrödinger,

I think that you have hit the nail on the head. It is true that the evasion using the arbitrarily large domain of cyclic variables to limit the value of Δp is very ingenious.[15] But an uncertainty relation interpreted that way does not appear to be very illuminating. The thing was invented for free particles, and it fits only that case in a natural way. Your claim that the concepts p,q will have to be given up, if they can only claim such a "shaky" meaning, seems to me to be fully justified. The Heisenberg-Bohr[16] tranquilizing philosophy—or religion?—is so delicately contrived that, for the time being, it provides a gentle pillow for the true believer from which he cannot very easily be aroused. So let him lie there.

But this religion has so damned little effect on me that, in spite of everything, I say

not: *E and* ν

but rather: *E or* ν;

and indeed: *not* ν, but rather *E* (it is ultimately real). But I cannot make head or tail of it mathematically. My brain is also

too worn out by this time. If you would give me the pleasure
of a visit from you again sometime it would be good of you
and very fine for me.

Best regards from
A. Einstein

15

Schrödinger to Einstein

7, Sentier des Lapins
LaPanne, Belgium
19 July 1939

Dear Einstein,

. . . .

A few months ago a Dutch newspaper carried a report which sounded comparatively intelligent that you have discovered something important about the connection between gravitation and matter waves. I would be terribly interested in that because I have really believed for a long time that the ψ-waves are to be identified with waves representing disturbances of the gravitational potential; not, of course, with those you studied first, but rather with ones that transport real mass, i.e. a non-vanishing T_{ik}. That is, I believe that one has to introduce matter into the abstract general theory of relativity, which contains the T_{ik} only as "asylum ignorantiae" (to use your own expression), not as mass points or something like that, but rather, shall we say, as quantized gravitational waves. I have done a good many calculations on this point but have found out very little, except that § 13.7 of Eddington's book "Protons and Electrons," which had fascinated me very much, is false. But it is unfortunately not very hard to find major errors in this ingenious book.

It's a shame that I had to fill so much of this letter with

uninteresting personal things about myself, but it is really so terribly hard to write, (I mean about such things as those just above).

If this letter reaches you on your sailboat I wish you much rest and enjoyment there. I am wonderfully well off here on the charming Belgian shore with these delightful people, happy as children. If one could only be somewhat more light-hearted and could think less about what is to become of oneself. Vacations are fine, but a vacation for which one cannot perceive a definite end is a peculiar thing.

> Best regards from
> Yours sincerely,
> *Schrödinger*

16

Einstein to Schrödinger

Peconic,[17] 9 VIII 1939

Dear Schrödinger,

. . . .

Now to physics. I am as convinced as ever that the wave representation of matter is an incomplete representation of the state of affairs, no matter how practically useful it has proved itself to be. The prettiest way to show this is by your example with the cat[18] (radioactive decay with an explosion coupled to it.) At a fixed time parts of the ψ-function correspond to the cat being alive and other parts to the cat being pulverized.

If one attempts to interpret the ψ-function as a complete description of a state, independent of whether or not it is observed, then this means that at the time in question the cat is neither alive nor pulverized. But one or the other situation would be realized by making an observation.

If one rejects this interpretation then one must assume that the ψ-function does not express the real situation but rather that it expresses the contents of our knowledge of the situation. This is Born's interpretation[19] which most theorists today probably share. But then the laws of nature that one can formulate do not apply to the change with time of something that exists, but rather to the time variation of the content of our legitimate expectations.

Both points of view are logically unobjectionable; but I cannot believe that either of these viewpoints will finally be established.

There is also the mystic, who forbids, as being unscientific, an inquiry about something that exists independently of whether or not it is observed, i.e. the question as to whether or not the cat is alive at a particular instant before an observation is made (Bohr). Then both interpretations fuse into a gentle fog, in which I feel no better than I do in either of the previously mentioned interpretations, which do take a position with respect to the concept of reality.

I am as convinced as ever that this most remarkable situation has come about because we have not yet achieved a complete description of the actual state of affairs.

Of course I admit that such a complete description would not be observable in its entirety in the individual case, but from a rational point of view one also could not require this.

I write this to you, not with any illusions that I will convince you, but with the sole intention of letting you understand my point of view, which has driven me into deep solitude. I have also brought it to the point of a real mathematical theory, whose testing, however, is naturally very difficult.

Best regards from

Yours,

A. Einstein

17

Schrödinger to Einstein

Innsbruck, Innrain 55
18 November 1950

Dear Einstein,

It seems to me that the concept of probability is terribly mishandled these days. Probability surely has as its substance a statement as to whether something is or is *not* the case—an uncertain statement, to be sure. But nevertheless it has meaning only if one is indeed convinced that the something in question quite definitely either *is* or *is not* the case. A probabilistic assertion presupposes the full reality of its subject. No reasonable person would express a conjecture as to whether Caesar rolled a five with his dice at the Rubicon. But the quantum mechanics people sometimes act as if probabilistic statements were to be applied *just* to events whose reality is vague.

The conception of a world that really exists is based on there being a far-reaching common experience of many individuals, in fact of all individuals who come into the same or a similar situation with respect to the object concerned. Perhaps instead of "common experience" one should say "experiences that can be transformed into each other in a simple way." This proper basis of reality is set aside as trivial by the positivists when they always want to speak only in the form: if "I" make a measurement then "I" "find" this or that. (And that is to be the only reality.)

It seems to me that what I call the construction of an external world that really exists is identical with what you call the describability of the individual situation that occurs only once—different as the phrasing may be. For it is just because they prohibit our asking what really "is," that is, which state of affairs really occurs in the individual case, that the positivists succeed in making us settle for a kind of collective description. They accuse us of metaphysical heresy if we want to adhere to this "reality." That should be countered by saying that the metaphysical significance of this reality does not matter to us at all. It comes about for us as, so to speak, the intersection pattern of the determinations of many—indeed of all conceivable—individual observers. It is a condensation of their findings for economy of thought, which would fall apart without any connections if we wanted to give up this mode of thought before we have found an equivalent that at least yields the same thing. The present quantum mechanics supplies no equivalent. It is not conscious of the problem at all; it passes it by with blithe disinterest.

It is probably justified in requiring a transformation of the image of the real world as it has been constructed in the last 300 years, since the re-awakening of physics, based on the discovery of Galileo and Newton that bodies determine each other's *accelerations*. That was taken into account in that we interpreted the velocity as well as the position as instantaneous properties of anything real. That worked for a while. And now it seems to work no longer. One must therefore go back 300 years and reflect on how one could have proceeded differently at that time, and how the whole

subsequent development would then be modified. No wonder
that puts us into boundless confusion!

Warmest regards!
Yours,
E. Schrödinger

18

Einstein to Schrödinger

_____

22 XII 1950

Dear Schrödinger,

You are the only contemporary physicist, besides Laue, who sees that one cannot get around the assumption of reality—if only one is honest. Most of them simply do not see what sort of risky game they are playing with reality—reality as something independent of what is experimentally established. They somehow believe that the quantum theory provides a description of reality, and even a *complete* description; this interpretation is, however, refuted, most elegantly by your system of radioactive atom + Geiger counter + amplifier + charge of gun powder + cat in a box, in which the ψ-function of the system contains the cat both alive and blown to bits. Is the state of the cat to be created only when a physicist investigates the situation at some definite time? Nobody really doubts that the presence or absence of the cat is something independent of the act of observation. But then the description by means of the ψ-function is certainly incomplete, and there must be a more complete description. If one wants to consider the quantum theory as final (in principle), then one must believe that a more complete description would be useless because there would be no laws for it. If that were so then physics could only claim the interest of shopkeepers and engineers; the whole thing would be a wretched bungle.

You are completely right to emphasize that the complete description cannot be built on the concept of acceleration, nor, it seems to me, can it be built on the particle concept. Only one of the tools of our trade remains—the field concept, but God knows whether this will stand firm. I think it is worthwhile to hold on to this, i.e. the continuum, as long as one has no really sound arguments against it.

But it seems certain to me that the fundamentally statistical character of the theory is simply a consequence of the incompleteness of the description. This says nothing about the deterministic character of the theory; that is a thoroughly nebulous concept anyway, so long as one does not know how much has to be given in order to determine the initial state ("cut").

It is rather rough to see that we are still in the stage of our swaddling clothes, and it is not surprising that the fellows struggle against admitting it (even to themselves).

Best regards!
Yours,
A. Einstein

Lorentz & Schrödinger

H. A. Lorentz

19

Lorentz to Schrödinger

Haarlem
27 May 1926

Dear Colleague,

I am finally getting around to answering your letter and to thanking you very much for kindly sending me the proof sheets of your three articles, all of which I have in fact received. Reading these has been a real pleasure to me. Of course the time for a final judgment has not come yet, and there are still many difficulties, it seems to me, about which I shall get to speak immediately. But even if it should turn out that a satisfactory solution cannot be reached in this way, one would still admire the sagacity that shows forth from your considerations, and one would still venture to hope that your efforts will contribute in a fundamental way to penetrating these mysterious matters.

I was particularly pleased with the way in which you really construct the appropriate matrices and show that these satisfy the equations of motion. This dispels a misgiving that the works of Heisenberg, Born, and Jordan, as well as Pauli's, had inspired in me: namely, that I could not see clearly that in the case of the H-atom, for example, a solution of the equations of motion can really be specified. With your clever observation that the operators q and $\frac{\partial}{\partial q}$ commute or do not commute with each other in a similar way to the q and p in the matrix calculation, I began to see the point. In spite of everything it remains a marvel that equations in which the q's and p's originally signified coordinates and momenta, can be satisfied when one interprets these symbols as things that have quite another meaning, and only remotely recall those coordinates and momenta. If I had to choose now between your wave mechanics and the matrix mechanics, I would give the preference to the former, because of its greater intuitive clarity, so long as one only has to deal with the three coordinates x, y, z. If, however, there are more degrees of freedom, then I cannot interpret the waves and vibrations physically, and I must therefore decide in favor of matrix mechanics. But your way of thinking has the advantage for this case too that it brings us closer to the real solution of the equations; the eigenvalue problem is the same in principle for a higher dimensional q-space as it is for a three dimensional space.

There is another point in addition where your methods seem to me to be superior. Experiment acquaints us with situations in which an atom persists in one of its stationary states for a certain time, and we often have to deal with quite definite transitions from one such state to another. Therefore we need to be able to represent these stationary states, every individual one of them, and to investigate them theoretically. Now a matrix is the summary of all possible transitions and it cannot at all be analyzed into pieces. In your theory, on the other hand, each of the states corresponding to the various eigenvalues E plays its own role.

Now permit me to make several comments in which, however, you probably will not find much new.

1. In your wave equation (I limit myself to the H-atom)

$$\Delta\psi + \frac{8\pi^2 m}{b^2}\left(E + \frac{e^2}{r}\right)\psi = 0 \qquad (1)$$

E is a constant independent of the coordinates; there are as many wave problems as there are energy values E, and of course the eigenvalues E are to be particularly considered here since only for these can the boundary conditions be satisfied. Your calculation of the eigenvalues* shows that one must understand E to be the energy of the electron, in the sense that the energy is set equal to zero when the electron is at rest at an infinite distance from the nucleus. Putting it another way, $E + \frac{e^2}{r}$ at any point x, y, z is the kinetic energy that the electron would have at that point for the prescribed value of E. This kinetic energy corresponds to the velocity

$$u = \sqrt{\frac{2}{m}\left(E + \frac{e^2}{r}\right)} \qquad (2)$$

2. Since Equation (1) contains no time derivative one can only derive from it the wave length at a definite point; one has, namely,

$$\frac{1}{\lambda} = \frac{1}{b}\sqrt{2m\left(E + \frac{e^2}{r}\right)} \qquad (3)$$

varying from point to point.

The velocity of propagation, w, of the waves, and the frequency, ν, related to it by the equation

---

* It is very beautiful that you were able to carry out this calculation and that in doing so you arrived at the values required by the Balmer formula.

$$w = v\lambda \tag{4}$$

cannot be derived from (1). A certain amount of arbitrariness remains here.

Now it is one of the basic ideas of your theory (and a very beautiful one) that the velocity, u, of the electron should be equal to the "group velocity." This requires the relationship

$$\frac{1}{u} = \frac{d}{dv}\left(\frac{v}{w}\right) \tag{5}$$

and if one takes this into consideration one can also determine v and w.

Concerning equation (5) it is to be observed first, that we want to consider v, ω, and u as all positive, and second, that at a definite point λ, u (and w) can vary with v, as follows from (2) and (3), because these quantities are somehow related to E. In carrying out the differentiation with respect to v that appears in (5) one must, however, abandon the eigenvalues E. There does not seem to be anything against this; one can very readily imagine states (travelling waves) which do indeed satisfy the wave equation, but do not satisfy all boundary conditions.

From (4) and (5) it follows that

$$\frac{1}{u} = \frac{d}{dv}\left(\frac{1}{\lambda}\right)$$

and therefore

$$\sqrt{\frac{m}{2\left(E + \frac{e^2}{r}\right)}} = \frac{1}{b}\frac{d}{dv}\sqrt{2m\left(E + \frac{e^2}{r}\right)}$$

and

$$v = \frac{1}{b}\left(E + \frac{e^2}{r} + \text{const}\right)$$

Since "const." means independent of E, we can set the constant equal to $E_0 - \frac{e^2}{r}$, where E_0 is not only independent of E but also of x, y, z. Thus,

$$v = \frac{1}{b}\,(E_0 + E) \tag{6}$$

By this means the condition that the frequency be equal at all points of the field is satisfied. Further, from (3) and (4),*

$$w = \frac{E_0 + E}{\sqrt{2m\left(E + \frac{e^2}{r}\right)}} \tag{7}$$

3. Your conjecture that the transformation which our dynamics will have to undergo will be similar to the transition from ray optics to wave optics sounds very tempting, but I have some doubts about it.

If I have understood you correctly, then a "particle," an electron for example, would be comparable to a wave packet which moves with the group velocity.

---

* If $E_0 + E$ is negative one can set

$$v = -\frac{1}{b}\,(E_0 + E), \quad w = -\frac{E_0 + E}{\sqrt{2m\left(E + \frac{e^2}{r}\right)}}$$

(both positive quantities), but equation (5) will then be satisfied by

$$u = -\sqrt{\frac{2}{m}\left(E + \frac{e^2}{r}\right)}$$

The wave velocity w and the group velocity u would have opposite directions in this case.

But a wave packet can never stay together and remain confined to a small volume in the long run. The slightest dispersion in the medium will pull it apart in the direction of propagation, and even without that dispersion it will always spread more and more in the transverse direction. Because of this unavoidable blurring a wave packet does not seem to me to be very suitable for representing things to which we want to ascribe a rather permanent individual existence.

As you yourself remark, the blurring in question is far advanced in the field of the H-atom. A wave packet can hold together for some time only if its dimensions are large compared to the wave length. Since, however, the wave length determined by (3) is of the order of magnitude of the Bohr elliptic orbit, there can be no question of having a wave packet that is small compared to the dimensions of such an ellipse and which is moving along this line.

Naturally, if you assign a large positive value to the constant E in (6) and (2), (one can think of $E = mc^2$), you can reach an arbitrarily high frequency ν with correspondingly large propagation velocity w, but you cannot change the wave length given by (3) at all.*

4. If we decide to dissolve the electron completely, so to

---

* If one puts $E = mc^2$ and $m = \dfrac{2}{3}\dfrac{e^2}{c^2 R}$ according to the usual formula (R is the radius of the electron), and if one also understands E to be energy in a circular Bohr orbit of radius r, so that $E = -\dfrac{1}{2}\dfrac{e^2}{r}$ then $w = c\left[\sqrt{\dfrac{2}{3}\dfrac{r}{R}}-\sqrt{\dfrac{3}{8}\dfrac{R}{r}}\right]$. Since $r \gg R$, we will have $w \gg c$. There is, naturally, nothing against that since we are dealing here with something quite different from the usual propagation of electromagnetic waves.

speak, and to replace it by a system of waves this has both an advantage and a disadvantage.

The disadvantage, and it is indeed a serious one, is this: whatever we assume about the electron in the hydrogen atom we must also assume for all electrons in all atoms; we must replace them all by systems of waves. But then how am I to understand the phenomena of photoelectricity and the emission of electrons from heated metals? The particles appear here quite clearly and without alteration; once dissolved, how could they condense again?

I do not mean to say by this that there cannot be many metamorphoses in the interior of atoms. If one wants to imagine that electrons are not always little planets that circle about the nucleus, and if one can accomplish something by such an idea, then I have nothing against it. But if we take a wave packet as model of the electron, then by doing so we block the way to restoring matters. Because it is indeed asking a lot to require that a wave packet should condense itself again once it has lost its shape.

The advantage that I spoke of consists of the following: if the electron continues to persist in a circular or elliptic orbit, one would then expect that in the wave equation, (1), (I am considering a point at which the electron is not located), there will appear not only the term e^2/r that depends on the field of the nucleus, but also a similar term that refers to the electric field of the electron. One field is as good as the other and they are of the same order of magnitude. But if equation (1) is changed this way the calculation of the eigenvalues of E would break down and would give rise to unspeakable complications. If the electron as such is no longer there then one can more readily be satisfied that only the term depending on the nuclear charge appears in the equation.

5. We will now replace Bohr's stationary states with energies E_1, E_2, etc. by "stationary wave systems" with frequencies

$$\nu_1 = \frac{1}{b}(E_0 + E_1), \quad \nu_2 = \frac{1}{b}(E_0 + E_2) \quad \text{etc.} \tag{8}$$

By giving the term E_0 a large positive value you can make these fundamental frequencies so high that they cannot be observed at all. (You can also assume that they are incapable of radiating, i.e. that there is no connection at all between the field which consists of the corresponding system of waves and the ordinary electromagnetic field, even though they both fill the same volume.) The observed radiations have the frequencies

$$\nu_i - \nu_k = \frac{1}{b}(E_i - E_k)$$

and the question arises as to how to account for this. Two ways suggest themselves to us—beats and combination tones.

There is not much to be said about the first. Let us suppose that we knew the fundamental equations from which the wave equation (1) results; I mean the true "equations of motion" which do not contain E at all, but contain time derivatives instead. If these fundamental equations are also linear then the superposition of two solutions, $\psi_1 = a_1 \cos(2\pi\nu_1 t + b_1)$ and $\psi_2 = a_2 \cos(2\pi\nu_2 t + b_2)$ will lead to beats; no instrument (resonator, grating) whose operation is completely determined by linear equations would respond to these beats as it would to vibrations of frequency $v_1 - v_2$. One can always imagine that somehow or other, although the process remains obscure for the present, a vibration takes place with the emission of radiation whose period corresponds to the frequency of the intensity maxima.

We can examine the origin of combination tones in somewhat more detail. To begin with, it is necessary that the fundamental equation be non-linear, but that is also sufficient. If, for example, a fundamental equation contains a term involving ψ^2, and if the vibrations that denoted by ψ_1 and ψ_2 are present at the same time, then as a consequence a term of the form

$$2\,\psi_1\,\psi_2 = a_1\,a_2\,\cos\left[2\pi(v_1 - v_2)t + b_1 - b_2\right] + \qquad (9)$$
$$+ a_1\,a_2\,\cos\left[2\pi(v_1 - v_2)t + b_1 + b_2\right]$$

will appear, where the first quantity just represents the difference tone. In order to understand quite clearly how this leads to radiation, however, one would have to take account of the connection between the vibrating system and the electromagnetic field. As far as the term denoting a sum in (9) is concerned one can assume that it cannot be made observable because of its high frequency $v_1 + v_2$.

In addition one can also understand absorption pretty well if one uses combination tones, which would be difficult to manage if one wanted to reduce optical phenomena to beats.

Let us suppose that the first vibrational state, $\psi_1 = a_1$ cos $(2\pi v_1 t + b_1)$ is already present in the atom and that now a force with frequency $v_2 - v_1$ acts on it (incident light). This can excite vibrations like

$$\psi' = a' \cos \left[2 \pi \left(v_2 - v_1 \right) t + b' \right]$$

(provided that the resonance is not *strong*). As a result the quantity

$$2 \, \psi_1 \, \psi' = a_1 \, a' \, \cos[2\pi v_2 t + b_1 + b'] +$$
$$+ \, a_1 \, a' \, \cos[2\pi(2v_1 - v_2)t + b_1 - b']$$

will appear in the term containing ψ^2 in the fundamental equation, and one can consider both of its parts as expressions for certain forces that excite vibrations of frequencies v_2 and $2v_1 - v_2$. The first of these, because its frequency coincides with the second characteristic vibration, can set the system into sympathetic oscillation (in this proper mode), and a part of the energy of the incident light is finally used for this. The force whose frequency is $2v_1 - v_2$ can remain ineffective because it corresponds to none of the characteristic vibrations of the system.

Naturally one could possibly try to pursue this kind of approach further.

What I do not like very much about this interpretation of radiation as produced by sum and difference oscillations is that the radiation is considered to be something of secondary importance, as something that depends on terms in the fundamental equations that one even neglects in first approximations (in deriving the wave equation (1).) Is it not really much simpler to hold onto Bohr's stationary states and then perhaps to assume that a Planck oscillator of frequency $(v_2 - v_1)$ is present, (the atom could turn into one), and finally that this absorbs the energy $h(v_2 - v_1)$ in a quantum jump $2 \rightarrow 1$ and then it calmly radiates?

6. Perhaps I may add that many years ago, when the laws of spectra were not yet known, my compatriot V. A. Julius[20] observed that in spectra containing many lines there are many pairs of lines for which Δv is almost the same. A probabilistic calculation, (similar to the one which served to prove that double stars are not accidental apparent approaches), then showed him that the number of differences Δv that differed from each other by less than a definite quantity ε, is much larger than one should expect according to the laws of chance. After he had shown the reality of the equations $\Delta v = \Delta' v = \Delta'' v = \ldots$ this way he arrived at the idea that many spectral lines might originate in sum and difference oscillations.

Rayleigh later made the observation that perhaps one could consider the simple appearance of the *first* power of the frequency in the spectral formulas (while dynamical laws lead rather to v^2) as an indication of kinematical relations.

After all these efforts I felt it to be a real simplification when Bohr showed that every frequency that is radiated corresponds to a definite energy difference, whereby the general structure of the spectral formulas immediately becomes clear. So I have lost my taste for explanations by means of sum and difference oscillations to some extent, but I can certainly reacquire it if your theory succeeds in other respects.

7. I find a real difficulty, as far as the combination oscillations are concerned, in the energy relationships. As far as the energy of the stationary wave system is concerned

one can make any arbitrary assumption to begin with, since one can dispose of the amplitudes freely, even if one has already accepted Eq. (6) for the frequency. However, it seems obvious that if one replaces Bohr's stationary states by stationary wave systems, one should assume certain definite energy differences between these. The fact that definite amounts of energy are necessary (in electron bombardment) to call forth definite radiation phenomena shows that the "energy levels" really exist, and if we no longer have the revolving electrons we have to look for the definite energy values in the individual stationary wave systems. The simplest thing would be to ascribe to these the energy values

$$E_0 + E_1, E_0 + E_2, E_0 + E_3, \text{ etc.} \tag{10}$$

Here $E_1, E_2, E_3, \ldots$ are the Bohr energy values (or also the eigenvalues of the wave equation), while E_0 is identified with the E_0 in (6), or if one prefers, it can be considered as different from the latter. In any case one probably has grounds for adding to the Bohr energy values a contribution, a positive one, E_0, which is the same for all wave systems. The values E_1, E_2, etc. are certainly negative and it is natural to represent the energy of a system of waves as a positive quantity.

If it is now assumed that the wave systems can exist with *only* the energy values (10) (so that they can have only certain prescribed amplitudes), a difficulty arises.

Let us suppose that state 1 is the "natural" one, the state of the atom when it is left to itself and the one that corresponds to the lowest energy, and furthermore let us ask that radiation of frequency $v_3 - v_2 = \dfrac{1}{b}(E_3 - E_2)$

be produced. According to Bohr we must first bring the atom into energy level 3, and we must therefore provide it with energy $E_3 - E_1$ (by electronic collision, say). The measurements are in accord with this. According to the new theory, however, we have to realize *both* states 2 and 3, since the required radiation presupposes the simultaneous existence of both. The *energy must then become* $E_0 + E_2 + E_0 + E_3$ while it was originally $E_0 + E_1$. If we assume that the first state of oscillation 1 disappears in the electronic collision, we find that the energy that has to be provided is $E_0 + E_2 + E_3 - E_1$ which is hardly to be reconciled with the observations.

One would naturally be able to escape this difficulty by assuming that the individual vibration states need not have quite the energies given by (10), but then what becomes of the energy levels?

Furthermore, according to Bohr just the energy $E_3 - E_2$ is radiated in the transition $3 \rightarrow 2$; state 3 disappears and is replaced by state 2. Can one imagine that just the energy $E_3 - E_2$ will be emitted in the radiation brought about by the difference oscillation, and then what becomes of the energy of both wave systems? Similar questions, which I need not go into, arise if the inverse process, absorption, is considered.

In conclusion I might say that one can feel that it is unsatisfactory in Bohr's theory that the frequencies emitted are completely distinct from the frequencies of the periodic motions that really take place. It is fine that in your theory both kinds of frequencies are brought into a much simpler connection with each other, (namely $v_{emitted} = v_2 - v_1$, where

v_1 and v_2 are "internal" frequencies); nevertheless it is not easy to understand this connection.

I should be very glad if you would write me sometime what you think about what is said above. Meanwhile please excuse me if perhaps I have not always correctly understood your meaning.

With kind regards
Yours faithfully,
H. A. Lorentz

20

Schrödinger to Lorentz

Zürich

6 June 1926

My dear Professor Lorentz,

You have rendered me the extraordinary honor of subjecting the train of thought in my latest papers to a profound analysis and criticism on eleven closely written pages. I cannot find words with which to thank you sufficiently for this precious gift that you have thereby made to me; I am deeply distressed that I have made such excessive demands on your time in this way. My thanks consist of—continuing these demands; but at least only for *reading*, and you have even given me permission to inform you about my attitude to the exceptionally interesting and important new viewpoints which your letter opens up. Please allow me to do this without directly answering the individual suggestions and doubts point by point; it is hardly likely that you still remember or have a copy of these in order. Also, much that I should like to say refers to several places in your letter.

1. You mention the difficulty of projecting the waves in q-space, when there are more than three coordinates, into ordinary three dimensional space and of interpreting them physically there. I have been very sensitive to this difficulty for a long time but believe that I have

now overcome it. I believe, (and I have worked it out at the end of the third article), that the physical meaning belongs not to the quantity itself but rather to a *quadratic* function of it. *There* I chose the real part of $\psi\,\bar{\psi}$, where ψ is taken to be complex in the obvious way (for criticism, see below) and the bar denotes the complex conjugate. *Now* I want to choose more simply $\psi\,\bar{\psi}$, that is, the square of the absolute value of the quantity ψ. If we now have to deal with N particles, then $\psi\,\bar{\psi}$ (just as ψ itself) is a function of 3N variables or, as I want to say, of N three dimensional spaces, $R_1, R_2, \ldots, R_N$. Now first let R_1 be identified with the real space and integrate $\psi\,\bar{\psi}$ over $R_2, \ldots, R_N$; second, identify R_2 with the real space and integrate over $R_1, R_3, \ldots, R_N$; and so on. The N individual results are to be added after they have been multiplied by certain constants which characterize the particles, (their charges, according to the former theory). I consider the result to be the electric charge density in real space. In this manner one obtains for an atom with many electrons exactly what Born-Heisenberg-Jordan designate as the transition probability, with the new and plausible meaning "component of the electric moment," (strictly speaking *that* partial moment which oscillates with the *emission* frequency in question).

What is unpleasant here, and indeed directly to be objected to, is the use of complex numbers. ψ is surely fundamentally a real function and therefore in Eq. (35) of my third paper

$$\psi = \sum_k c_k u_k\,(x)\, e^{\dfrac{2\pi i E_k t}{h}} \tag{35}$$

I should be good and write a cosine instead of the exponential, and ask myself: is it possible in addition to define the imaginary part unambiguously *without reference to the whole behavior of the quantity in time*, but rather referring only to the real quantity itself and its time and space derivatives *at the point in question.* This actually can be done, at least for ψ. I write the "wave equation" for the sake of brevity in the form

$$L[u] + Eu = 0 \tag{1}$$

where $L[\ ]$ means a certain differential operator. Furthermore let ψr be the *real* wave function, the only one known originally, therefore the real part of ψ, whose imaginary part is to be defined in addition. That can be done this way:

$$\dot{\psi} = \dot{\psi}r - \frac{2\pi i}{b}L[\psi r] \tag{2}$$

By this method, therefore, the magnitude of $\dot{\psi}$ is in any case represented by the space and time derivatives of the real quantity ψr, independent of the complex representation; so that one does not get into difficulty even if there is a ψr that does not correspond to a stationary superposition of proper oscillations. Now to be sure one has only $\dot{\psi}$ and integration with respect to time would involve an undetermined additive purely imaginary function of the coordinates. I do not know yet whether this can be fixed in a rational way. Probably there is nothing that prevents us in practice from replacing ψ by $\dot{\psi}$ throughout the argument cited first, since all eigenvalues are really almost equally large because of the large additive constant that they contain, and of which you also speak. If this constant is fixed to be mc^2 (or an integer multiple of it), which is practically inevitable, then the *differences* in the eigenvalues become very small compared to the eigenvalues themselves, of the order of the relativistic correction.

2. You often touch on the point that the "wave equation" (1) is still not the fundamental equation of the problem, because it no longer contains any time derivatives but has instead the integration constant E. Also the equation does not hold in general but only for such solutions u that depend on time through the factor { }.[21] But the latter is equivalent to

$$\ddot{u} = -\frac{4\pi^2}{b^2} E^2 u$$

One can eliminate E between (1) and (2)[22] and one obtains

$$L\, L[u] + \frac{b^2}{4\pi^2} \ddot{u} = 0 \tag{3}$$

This might well be the *general wave equation* which *no longer* contains the integration constant E but contains time derivatives instead. It is of completely the same type as the equation for a vibrating *plate* (which contains the repeated Laplacian operator), and no longer of the simple type of the vibrating membrane. It took me a terribly long time to discover this simple fact. Of course one can now go backwards again from equation (3) by trying the form

$$u \sim e\, \frac{2\pi i E t}{h}$$

and trying to split (3) according to the pattern

$$(L[\,] - E)(L[\,] + E)u = 0$$
$$L[u] - Eu = 0 \quad \text{or} \quad L[u] + Eu = 0$$

as in the case of the vibrating plate. That one obtains *all* solutions this way has to be proved afterwards by investigation of the completeness of the system of functions that is found. (Naturally it makes no difference that *two* equations with different signs for E result, because E is an undetermined constant, still to be determined; one does not therefore obtain any *new* solutions in addition.)

3. Allow me to send you, in an enclosure, a copy of a short note in which something is carried through for the simple case of the oscillator which is also an urgent requirement for all more complicated cases, where however it encounters great computational difficulties. (It would be nicest if it could be carried through *in general*, but for the present that is hopeless.) It is a question of really establishing the wave groups (or wave packets) which mediate the transition to macroscopic mechanics when one goes to large quantum numbers. You see from the text of the note, which was written *before* I received your letter, how much I too was concerned about the "staying together" of these wave packets. I am very fortunate that now I can at least point to a simple *example* where, contrary to all reasonable conjectures, it still proves right.

I hope that this is so, in any event for all those cases where ordinary mechanics speaks of *quasi-periodic* motions. Let us accept this as secured or conceded for once; there still always remains the difficulty of the completely *free* electron in a complete field-free space. Would you consider it a very weighty objection against the theory if it were to turn out that the electron is incapable of existing in a completely field-free space? Or perhaps even that "free" electrons do not permanently keep their identities at all in the usual sense? That speaking of individual electrons in a bundle of cathode rays perhaps means only that the bundle has a certain "granular" structure, in just the same way that many phenomena have made this seem plausible for a bundle of light rays, where in *both* cases neither a pure wave description nor a pure particle description exactly reaches the truth, but rather something in between that we have not yet adequately achieved.

4. I should like to add a few more remarks to the considerations in the enclosed note, the most important of which seems to me to be this: one should *not* set the *individual* proper oscillations of the wave theory in parallel with the *individual* stationary orbits of the Bohr theory. For if one does that the transition from micromechanics to macromechanics by means of the correspondence principle is absolutely impossible. One can even see how for large quantum number (A >> 1) the individual Bohr orbits are built up by a superposition of very many proper oscillations which are relatively closely adjacent to one another. It would be possible for imposed couplings between the amplitudes and phases of adjacent proper oscillations to persist, perhaps in such a way that one obtains all possible states of the oscillator, while one allows the quantity A to assume all possible positive values, (the whole aggregate then must still be thought of as multiplied by $e - \frac{A^2}{2}$ so that the integral $\int_{-\infty}^{+\infty} \psi\bar{\psi}dx$ will be independent of A.) In the limiting case of very small A one obtains at first only the fundamental oscillation; with increasing A the higher oscillations are gradually excited and the center of gravity gradually shifts to higher and higher quantum numbers.

But for the present these are just chimeras; it could be completely different. In no case do I consider it correct to speak of the *energy of the individual proper oscillation*, measured perhaps by the *square of its amplitude*. In my view the latter has nothing to do with energy but rather with *charge*. The *only* property of the individual proper oscillation that has anything to do with energy is its *frequency*, I believe.

The question naturally arises: but why must I supply a quite definite amount of energy to the atom in order just to excite a definite proper oscillation? Now "supply a definite amount of energy" really means here either "bombard with electrons of definite velocity" or "irradiate with light of definite frequency." As far as the *latter* is concerned you will know better than I that a physicist of the old days would have opened wide his eyes and his mouth if someone had said to him: to irradiate with light of definite frequency "means" to supply a definite amount of energy. He would have looked for a very much more obvious explanation in *resonance*. The basis for the statement just made, which would be so hard to understand for the physicist of the old days, can be seen in the fact that light of a definite frequency is always capable of producing the same physical effects as electrons of a definite velocity. From the fact of this equivalence the opposite conclusion can, however, be drawn with the same inevitability, or lack of inevitability: the electron moving with a definite velocity must be a wave phenomenon whose frequency is that of the fight which is experimentally equivalent to it with regard to the excitation of resonance. I consider one conclusion to be as one-sided as the other, the truth lying somewhere in the middle.

5. You discuss the question of the explanation of radiation by means of beats or by means of difference tones in a very penetrating way that is also very instructive for me. I must frankly admit that up to now I have not made enough of a conceptual distinction between these two things. I was so extremely happy, first of all, to have arrived at a picture in which at least something or other really takes place with that frequency which we observe in the emitted light that, with the rushing breath of a hunted fugitive, I fell upon this something in the form in which it immediately offered itself, namely as the amplitudes periodically rising and falling with the beat frequency. By this I only meant: there is a *conceivable* mechanism by means of which these rising and falling amplitudes can excite light of *equal* frequency. The frequency discrepancy in the Bohr model, on the other hand, seems to me, (and has indeed seemed to me since 1914), to be something so *monstrous*, that I should like to characterize the excitation of light in this way as really almost *inconceivable*. Between the *alternatives* of beats or difference tones, however, I obviously declare myself for the latter. That of course only means that it is not at all necessary for everything to happen strictly *linearly*; otherwise the most beautiful beat frequency remains eternally ineffectual.

6. I am surprised that at one point in your letter you take strong offense at what you describe in the words "the radiation is considered to be something of secondary importance, as something that depends on terms in the fundamental equations that one even neglects in first

approximation (in deriving the wave equation)." In case I understand you correctly I must declare that I would consider the opposite to be a serious stumbling block if it were to be proposed.

And that would be because I believe that the significance of the radiation terms for atomic dynamics is grasped correctly, so far as orders of magnitude are concerned, by the older theories, and indeed it was not the Bohr theory that did this first, but rather the electron theory. The radiation terms play a thoroughly secondary role in both theories. The main force exerted by the self field on the electron in the theory of electrons is the inertial force. The radiation force appears first as the second term in a series expansion and is really always very small compared to the inertial force in the typical electronic motions that occur. In the Bohr model too the radiation reaction force is completely disregarded first of all, and the whole model is built up without it. It does not come in until afterwards and then through two things: first through the assumption of a *lack of sharpness* of the levels (determined just by the classical radiation force), and second, through the "electron jumps." The latter, to be sure, can no longer be compared *directly* to anything else with respect to order of magnitude because of their bizarre discontinuity. But since the *frequency* with which the jumps occur is still calculated from the radiation force by using the correspondence principle, it can be seen that it is to be put on the same level as the latter so far as order of magnitude is concerned. I am therefore quite satisfied that the wave mechanics seems to be in agreement with the older theories on this point insofar as the reaction of the radiation on the radiating system is insignificant enough so that it can be neglected to the first approximation in setting up its "equation of motion."

Through your discussion I have become absolutely certain that the addition of these terms must necessarily put an end to the linear character of the equations of motion, and I consider this an exceptionally important piece of knowledge.

7. From the wave equation itself and from the assumption about the group velocity you again derive in reverse the expression in Eq. (6) of my second paper for the wave velocity, from which I started:

$$u = \frac{E + E_0}{\sqrt{2m(E - V)}}$$

The constant E_0 is formally absent from my equations, yet I emphasized (p. 10 of that paper) that E and V individually are obviously determined only up to an additive constant. I took particular pleasure in calling special attention (at that very place) to the circumstance that the *wave length* is independent of this constant, just because the wave length determines the order of magnitude of the orbital dimensions at which quantum phenomena begin to appear.

Then, at a place later on, you emphasize that, just because of this unalterably fixed wave length, the dimensions of the electron are certainly of at least the same order of magnitude as the Bohr elliptic orbits of low quantum number, and that it is in no way possible to construct wave packets which revolve in these orbits and are small compared to the dimensions of the orbit. I do not know whether I am correct in reading an "unfortunately" between the lines. But I believe that the enclosed note shows you that, in any event, I never cherished this wish for the orbits of *low* quantum number. In my view *these* states are something that differs *toto genere* from electron orbits; not until the high quantum numbers does classical mechanics gradually assume its rights again, just as the diffraction image of a slit is gradually transformed into its shadow image if you slowly pull the sides of the slit apart.

8. In conclusion may I emphasize several serious difficulties of a fundamental nature in the matrix mechanics, (without any connection with your letter), which have gradually become clear to me and in which I see an advantage in the wave mechanics, quite apart from its intuitive clarity.

Most important of these is the *symmetrization of the Hamiltonian function*. I have spoken about this in considerable detail on page 14 of my third article. But what I had not clearly recognized yet at that time was and is that the rules set up for this purpose by Born, Jordan and Heisenberg are actually false if one applies them to generalized coordinates; they are correct only in Cartesian coordinates. That has turned out simply *empirically* in the calculations of Dirac and Pauli; that symmetrization is then chosen which leads to something reasonable. Heisenberg, in a summarizing article in the *Mathematische Annalen*,[23] has therefore decided to lay down the rule that the Hamiltonian must be taken over from the classical theory in *Cartesian* coordinates. In doing so he does not, however, explicitly retract the abovementioned strictly false generalization to arbitrary coordinates (given earlier in a paper with Born and Jordan in the *Zeitschrift für Physik*.) Furthermore there are still situations left that are completely undetermined, such as the symmetric or the asymmetric top, since here returning to Cartesian coordinates is not only cumbersome but even impossible so long as it has not been decided how "rigid connections" are to be interpreted in the new mechanics.

Wave mechanics, on the other hand, is directly applicable to arbitrary coordinates and allows the energy levels to be calculated without even having to know the connection between the general coordinates and Cartesian coordinates.

A second point is that wave mechanics always yields completely *determined* eigenvalues, apart perhaps from *one* additive constant (which is of no consequence in the energy *differences*). In matrix mechanics this seems to be very difficult at least, and I am not sure if at times there are not still some things that are indefinite in *principle*. Dirac (*Proc. Roy. Soc.*)[24] and Wentzel (*Zeitschrift fur Physik*)[25] calculate for pages and pages on the hydrogen atom, (Wentzel relativistically, too), and finally the only thing missing in the end result is just what one is really interested in, namely, whether the quantization is in "half integers" or "integers"!* Thus, Wentzel does indeed find "exactly the Sommerfeld fine structure formula" but for the reasons mentioned the result is completely worthless for comparison with experiment. The relativistic treatment by means of wave mechanics, which is just as simple, results unambiguously in half integral azimuthal and radial quanta, just like the classical treatment. (I did not publish the calculation at the time because this result just

---

* Each of the quantum integrals still contains an additive constant which remains undetermined. The only thing that can be concluded is that they are *progressively* larger by integral multiples of h. That is a serious deficiency and not an unimportant one like the additive energy constants.

showed me that something was still missing; that something is certainly Goudsmit and Uhlenbeck's idea.) As a side remark, Wentzel's procedure is *so* constituted that *if* he were to push through to the result, his result would probably be false,* because he takes the problem as two dimensional rather than three dimensional. That is not permitted, as I stressed in my second article, p. 32, and, because of the complete mathematical equivalence of wave mechanics and the Göttingen mechanics, it is also certainly not permitted in the latter. Wave mechanics also allows one to perceive clearly the reason for this, because a wave motion in two dimensions is obviously something completely different from a wave motion in three dimensions. In the Göttingen mechanics on the other hand one cannot really understand, so far as I can see, why the reduction of the problem by the use of an integral should be forbidden. At least the reason is not very evident or else it would not be generally used.

I am afraid, Professor Lorentz, that I have taken up a great deal of your time again with this long letter. But your criticism of my attempt—kind, penetrating, and yet, despite all your misgivings, well-meaning—allows me to hope that one or another of the ideas that you induced will be of interest to you. I am quite convinced that I have not been able to dispel all your misgivings; to tell the truth, I have more than enough of them myself, and in all these considerations I perceive no more than the first pale glimmerings of what I hope is the dawn of a more profound understanding.

I must also thank you very much for something else, and that is the charming picture with which you rewarded all those who demonstrated their respect for you (on the

---

* i.e. would not represent the correct assertion of the theory.

occasion of your festival day[26]) by what was, at least in my case, unfortunately a purely symbolic action. The delightful picture will always be a beautiful reminder of the days of unalloyed pleasure that I was allowed to spend under your guidance in Brussels two years ago.[27]

I beg you always to be assured of the sincere admiration
and respect of
Yours faithfully,
E. Schrödinger

21

Lorentz to Schrödinger

Haarlem

19 June 1926

Dear Colleague,

I read your last letter, for which many thanks, with lively interest, and it helped a great deal in making your interpretations clear to me. Now I see that the difficulties that I experienced turned partly on the fact that I had become very accustomed to the ideas of the current quantum theory, so that I could not immediately free myself from it sufficiently. That is why, for example, I objected that radiation appears as something "secondary" in your work.

You are quite right when you say that this is also the case in classical theory inasmuch as the term in the equation of motion of the electron that corresponds to the radiation resistance falls far short of the other terms, so that it can often be neglected in first approximation. But I was thinking of a quantum jump 2–1, in which (as I imagined along with Bohr) the definite finite quantity of energy $E_2 - E_1$ is radiated with frequency $v_{21} = \frac{E_2 - E_1}{b}$. Such transitions might occur rarely but the radiation is actually the main thing when each individual quantum jump does occur.* But if your

---

* In order to picture the process to some extent, I imagined that there is an oscillator of frequency v_{21} which absorbs the energy $E_2 - E_1$ and then quietly radiates it; or also that when the atom has energy E_2, it

interpretation (radiation as a difference tone) can be carried through successfully, and if we then no longer have to think of the radiation of just the quantity of energy $E_2 - E_1$ I will also be satisfied.

In this connection I was also very pleased with your remark about a moving electron's "capacity for exciting radiation." Here too I was thinking too much about the energy of the electron. If one can successfully explain the phenomena by connecting a definite frequency to the moving electron so that one is dealing with a resonance, it would be much more beautiful.

Meanwhile there are still many questions that arise here. Suppose we have a system with fundamental frequencies v_1 and v_2, and that

$$v_1 = \frac{E_0 + E_1}{b}, \quad v_2 = \frac{E_0 + E_2}{b} \tag{1}$$

where E_1 and E_2 are the (negative) energies that we ascribe to the atom in two stationary states (according to Bohr), while E_0 has a large positive value. One can now imagine that under the influence of an external irradiation whose frequency is $v_2 - v_1$ the system will be caused to emit light of this same frequency ("resonance with the difference tone"). But how is resonance with an electron to occur? In de Broglie's case (electron moving in a straight line) one must distinguish between the frequency in the interior of the electron and that of the waves that accompany the particle in its motion. I will keep to the first one here because I do not have a sufficiently clear idea of the waves in this case.

---

transforms itself into an oscillator v_{21} for a time, and the latter becomes a Bohr atom again when its energy has decreased from E_2 by radiating.

So far as the internal frequency is concerned, if this has the value ν_0 for an electron at rest, then when the electron moves with a velocity v its frequency will amount to

$$\nu_0\sqrt{1 - \frac{v^2}{c^2}} = \nu_0 - \nu_0\frac{v^2}{2c^2}$$

according to relativity theory. Probably one can hardly do anything else but take $\nu_0 = \frac{mc^2}{b}$. The frequency then becomes

$$\frac{mc^2 - \frac{1}{2}mv^2}{b}$$

According to experiment the electron can only cause radiation of frequency $\nu_2 - \nu_1$ if

$$\frac{1}{2}mv^2 = E_2 - E_1$$

so that the last expression becomes

$$\frac{mc^2 + E_1 - E_2}{b} \tag{2}$$

Now how can a system whose fundamental frequencies are given by (1) be caused to resonate so that it radiates $\nu_{21} = \frac{E_2 - E_1}{b}$ under the influence of a disturbance of frequency (2)? One does not understand it even if one puts $E_0 = mc^2$, which is the natural thing to do; and the matter is further complicated by the fact that the electron is flying through the electron,[28] so that it more quickly feels the oscillation field at different points with its rapid vibrations, so that probably something like a Doppler effect must still be taken into consideration.

You gave me a great deal of pleasure by sending me your note, "The continuous transition from micro- to macro-mechanics"[29] and as soon as I had read it my first thought was: one must be on the right track with a theory that can refute an objection in such a surprising and beautiful way. Unfortunately my joy immediately dimmed again; namely, I cannot comprehend how, e.g. in the case of the hydrogen atom, you can construct wave, packets that move like the electron; (I am now thinking of the *very high* Bohr orbits). The *short* waves required for doing this are not at your disposal. I already referred briefly to this point in my first letter and should now like to go somewhat further into it. Before that, however, permit me to communicate to you some calculations that were prompted by your note. Maybe the method that I used there can be applied in some case or other.

Since at present we hardly dare hope really to construct the wave packets in more complicated situations, I asked myself the question: if one *assumes* that there are wave groups that remain permanently confined to a small volume, can one then prove that they have to move in a field of force exactly as an electron would? This could naturally be proved immediately if the assertions of ordinary optics concerning propagation (light rays, group velocity) might be taken over to the present case. But one must be careful with this taking over; as you observe, optics talks about a continuous series of frequencies, but here we have only individual discrete frequencies. Your result already shows that in the case under consideration something else can be derived (and indeed more, namely a wave packet that really stays together permanently), than from the aforesaid optical theorems.

I tested the method first on the linear oscillator and then applied it to the H-atom.

In the original a calculation requiring 12 pages follows at this point whose result is that a wave packet does not remain intact on a high quantum number orbit in the hydrogen atom and hence cannot be used as a model of an electron.

This is the reason why it seems to me that in the present form of your theory you will be unable to construct wave packets that can represent electrons revolving in very high Bohr orbits. For we may surely take this much from classical optics:* a wave packet must include very many wave lengths. You had the advantage in your example of the linear oscillator of having arbitrarily short waves at your disposal.

In your letter you talk about having a certain quantity quadratic in ψ mean the electric charge density (and not perhaps an energy) where you imagine the electron to be "smeared out." I should just like to ask whether it would not be nice (and desirable) if $\int \rho \, d\tau$ were to be a constant, if we are to identify a quantity appearing in the equations as the charge density? That would hardly be allowed to prove right if $\rho = \psi \, \bar{\psi}$. Would it not be more natural to take ρ as having the value that I denoted by ε, and called the energy in the preceding calculation? $\int \varepsilon \, d\tau$ is indeed constant.

A second question: can you distinguish between positive and negative charge?

---

* It could probably also be derived from the equation of motion now under consideration (analogue of Huygens' principle).

One difficulty, which I already alluded to, is that the V which appears in the formulas (with the term $-\frac{e^2}{r}$) refers only to the field of the nucleus; can one confine himself to *this* potential if negative charge is also present, either continuously distributed in space or concentrated in an electron? If one alters the term $\frac{e^2}{r}$ one runs the risk of losing the correct eigenvalues for E.

These are all obscure points. On the other hand it is once again gratifying that by making $\psi\,\bar{\psi}$ responsible for the emission of radiation, (you could obtain the same result with any quadratic quantity), you already allow the difference tones and the radiated frequencies to appear without the need for any further assumptions (non-linearity of the equations).

If you will permit me I should like to conclude with a brief summary of what, in my opinion, can be said about your theory now, so far as it is developed and so far as it can be maintained; I am thinking particularly of the H-atom in this connection. In doing so I put aside the energy packets and also do not talk about the blurring or dissolving of the electron.

1) In the field of the nucleus there can exist oscillating wave states which belong to a definite equation of motion. Rules are given for deriving these from the equations of motion of an electron.

   The potential that appears in the equation of motion depends upon the nuclear charge. The charge of the electron has no influence on this potential.

2) The possible wave states have definite (very high) frequencies which are found by considering the boundary conditions (for r = o and r = ∞). At every point there are definite w and λ, depending on the point but independent of direction.

3) A quantity quadratic in ψ is made responsible for the emission of radiation. As soon as two of the states of motion already mentioned, with frequencies, v_1 and v_2, exist at the same time, this leads to the *radiated frequency* $v_2 - v_1$ (and to a frequency $v_2 + v_1$, which is *very* high and which we are allowed to [or want to] disregard).

Thus far nothing is said about the electron. But it must somehow or other take part in the proceedings as already follows from the fact that the spectrum of an atom is fundamentally changed by the loss of an electron. For that reason I shall still add the following.

4) For any of the states of oscillation mentioned above there are certain *specially distinguished* lines* characterized by the condition,

$$\delta\int\frac{ds}{w} = 0 \qquad (32)$$

for fixed end points, where ω is the velocity of propagation. The specially distinguished curves for the n-th state are precisely the n-quanta orbits of the electrons in the Bohr Theory.

Proof: one can replace (32) by

$$\delta\int\frac{ds}{\lambda} = 0 \qquad (33)$$

Now for the n-th state, which we want to consider, E is fixed at E_n, and in the wave equation

$$\Delta\psi + \frac{2m}{H^2}\left(E + \frac{e^2}{r}\right)\psi = 0 \qquad (34)$$

---

* I refer to them as lines and do not speak of "light rays" because there is no longer any question of the physical significance of the latter (limitation of a wide bundle of rays).

$E + \frac{e^2}{r}$ represents the kinetic energy $\frac{1}{2}mv^2$, that an electron with the total energy E_n would have at the point in question.[30] If one derives λ from (34) one obtains $\lambda \sim \frac{1}{v}$ (with a constant coefficient); thus (33) is transformed into the condition that $\int v \, ds = 0$, for prescribed E_n. But this is just the condition that determines the motion of the electron.

5) At the same time it can be seen that the specially distinguished curves are closed (ellipses or circles). Now they have the additional property that their circumference, expressed in wave lengths (I mean $\int \frac{ds}{\lambda}$) is an integer.*

Proof: From (34) we obtain for the wave length:

$$\frac{4\pi^2}{\lambda^3} = \frac{2m}{H^2} \frac{1}{2} mv^2 = \frac{m^2 v^2}{H^2}, \quad \lambda = \frac{2\pi H}{mv} = \frac{b}{mv}$$

Hence

$$\int \frac{ds}{\lambda} = \frac{1}{b} \int m \, v \, ds = \frac{2}{b} \int T \, dt = \frac{2}{\lambda} \Theta \, \overline{T}$$

if Θ is the time it takes for the electron to go around once in the orbit under consideration and T is the time average of the kinetic energy. But for motion in a Kepler ellipse the theorem

$$\overline{T} = -E$$

is valid if E is the energy, (where the potential energy is zero at infinity.) We must therefore calculate

---

* We can talk about this even without thinking of a propagation along the lines.

$$-\frac{2}{b} \ominus E_n$$

and can do this for a circular orbit, since the time of revolution $\ominus$ is the same for all n-quanta orbits, whether they are circles or ellipses. Now for a circular orbit of radius r_n,

$$E_n = -\frac{e^2}{2r_n}, \quad \ominus = \frac{2\pi r_n}{v_n}$$

so that our expression becomes

$$\frac{2\pi e^2}{b v_n}$$

Now, since v_n can be evaluated from the known formula $v_n = \frac{2\pi e^2}{nb}$, we obtain

$$\int \frac{ds}{\lambda} = n$$

6) For some reason or other* the electron can move only along the specially distinguished curves. In connection with this we remain somewhat uncertain as to what the electron will do if two of the states of oscillation exist at the same time.

As you see, what has just been said comes close to de Broglie's arguments. As compared to him you have made the advance of setting the wave states clearly before us, and that is an important step.

Nevertheless, if we have to give up wave packets and with

---

* Hard to say why. One could think here of de Broglie's interpretation: internal vibrations of the electron and agreement in phase between this and the accompanying wave.

them one of the basic ideas of your theory, the transformation of classical mechanics into a wave mechanics, something would be lost that would have been very beautiful. I should be very pleased if you could find a way out of this.

For the rest, I would be very satisfied if one could get as far with several other cases (relativistic correction, relative motion of the nucleus, Stark and Zeeman effects) as with the Balmer spectrum as summarized in 1–6 above.

> With kind regards,
> Yours faithfully,
> *H. A. Lorentz*

A Biography of Albert Einstein

––––––––––

ALBERT EINSTEIN (1879–1955) is among modern history's greatest and most influential minds. He authored more
than 450 scholarly works during his lifetime, and his advance-
ments in science—including the revolutionary Theory of
Relativity and E=mc², which described for the first time the
relationship between an object's mass and its energy—have
earned him renown as "the father of modern physics."

Born in Ulm, in southwest Germany, Einstein moved
to Munich with his family as an infant. As a child, Einstein
spoke so infrequently that his parents feared he had a learn-
ing disability. But despite difficulties with speech, he was
consistently a top student and showed an early aptitude for
mathematics and physics, which he later studied at the Swiss
Federal Institute of Technology in Zurich after renouncing
his German citizenship to avoid military service in 1896.

After graduation, Einstein married his college girlfriend,
Mileva Marić, and they had three children. He attended the
University of Zurich for his doctorate and worked at the patent office in Bern, a post he left in 1908 for a teaching position
at the University of Bern, followed by a number of profes-
sorships throughout Europe that ultimately led him back to
Germany in 1914. By this time, Einstein had already become
recognized throughout the world for his groundbreaking
papers on special relativity, the photoelectric effect, and the

relationship between energy and matter. He won the Nobel Prize in Physics in 1921.

In 1933, Einstein escaped Nazi Germany and immigrated to the United States with his second wife, Elsa Löwenthal, whom he had married in 1919. He accepted a position at Princeton University in New Jersey, where he stayed for the remainder of his life. At Princeton, Einstein dedicated himself to finding a unified field theory and played a key role in America's development of atomic weapons. He also campaigned for civil rights as a member of the NAACP and was an ardent supporter of Israel's Labor Zionist Movement.

Still, Einstein maintained a special affinity for his home-land. His connection to all things German and, in particular, to the scientific community in Berlin was probably the reason that throughout his years in America he so strongly valued his relationships with other German-speaking immigrants. He maintained a deep friendship with the founder of Philo-sophical Library, Dr. Dagobert D. Runes, who, like Einstein, was a humanist, a civil rights pioneer, and an admirer of Baruch Spinoza. Consequently, many of Albert Einstein's works were published by Philosophical Library.

At the time of Einstein's death in 1955, he was universally recognized as one of history's most brilliant and important scientists.

Einstein with friends Marcel Grossmann, Eugen Grossmann, and Gustav Geissler in the garden of the Grossmann home in Thalwil, Switzerland, around 1899. Einstein's discussions with Marcel about elliptic geometry provided one of the sparks that led to Einstein's development of the General Theory of Relativity.

Einstein with his first wife, Mileva Marić, and their son Hans Albert, in 1904. Their second son, Eduard, would be born six years later.

A twenty-six-year-old Einstein during the time he was employed at the Bern patent office, in 1905.

Paper silhouettes created by Einstein in 1919, the year of his marriage to his second wife, Elsa. The silhouettes depict, from left to right, himself, Elsa, and his stepdaughters Ilse and Margot.

Einstein lecturing in Vienna, Austria, in January of 1921, the same year he won the Nobel Prize in Physics. 1921 also marked the year of Einstein's first visit to New York City, followed by weeks of lectures at some of the East Coast's most prestigious universities.

Einstein with Elsa in Migdal, Israel, on February 12, 1923.

Einstein smoking a pipe on the porch of his home in Princeton, New Jersey, in 1938. He was a very ardent pipe smoker and treasured the ritual of selecting different tobaccos and preparing them to be smoked.

Einstein with his friends poet Itzik Feffer and actor Solomon Mikhoels, in 1943.

Einstein in his Princeton study on the day that he received his honorary degree from the Hebrew University of Jerusalem, in 1949.

Einstein receiving the honorary degree from Israel S. Wechsler while at his Princeton home in 1949.

*A portrait of Einstein at the Yeshiva University inauguration
dinner for the Albert Einstein College of Medicine, at Princeton
Inn on March 15, 1953.*

Notes

INTRODUCTION

1. E. Schrödinger, *Annalen der Physik* 79 (1926) p. 361.

2. M. Fierz and V. F. Weisskopf, editors, *Theoretical Physics in the Twentieth Century* (New York: Interscience Publishers Inc., 1960) p. 22.

3. W. Heisenberg, *Zeitschrift für Physik* 33 (1925) p. 879. M. Born and P. Jordan, *Zeitschrift für Physik* 34 (1925) p. 858. M. Bom, W. Heisenberg, and P. Jordan, *Zeitschrift für Physik* 35 (1926) p. 557.

4. E. Schrödinger, Annalen der Physik 79 (1926) p. 734.

5. Reference 2, p. 44.

6. W. Pauli, Editor, *Niels Bohr and the Development of Physics* (New York and London: Pergamon Press, 1955) p. 14.

7. See the papers collected in E. Schrödinger, *What is Life? and Other Scientific Essays* (Garden City, New York: Doubleday and Company, 1956).

8. *Louis de Broglie, Physicien et Penseur* (Paris: Editions Albin Michel, 1953) p. 165.

9. See M. J. Klein, *Einstein and the Wave-Particle Duality* in *The Natural Philosopher* (New York: Blaisdell Publishing Company, 1964) III pp. 1–49.

10. E. Schrödinger, Reference 7, pp. 132–133.

LETTERS 1-8: SCHRÖDINGER & PLANCK

1. Schrödinger's early works on wave mechanics are collected in his book *Collected Papers on Wave Mechanics* (Glasgow: Blackie and Son, 1928). The first group appeared in the *Annalen der Physik* (4) 79 (1926) pp. 361, 489, 734; 80 (1926) p. 437; 81 (1926) p. 109.

2. Schrödinger had inadvertently spelled the name of the mathematician Jacobi with a k.

3. This probably refers to Planck's lecture, *Physikalische Gesetzlichkeit im Lichte neuer Forschung*, delivered on 14 February 1926 in Dusseldorf. *Naturwissenschaften* 14 (1926), p. 249. Reprinted in Max Planck, *Physikalische Abhandlungen und Vorträge* (Braunschweig: Friedr. Vieweg und Sohn, 1958) Vol. III, p. 159.

4. Grüneisen was at that time President of the Berlin branch of the German Physical Society.

5. See letter No. 19. It was actually 11 pages; see letter No. 20.

6. Schrödinger's lecture, entitled "Foundations of an atomism based on the theory of waves," was given on 16 July 1926 in Berlin with W. Nernst in the chair. A similar lecture was delivered on 23 July 1926 to the Bavarian branch of the Society with R. Emden in the chair.

7. M. Planck, Verhandlungen der Deutschen Physikalischen Gesellschaft *13* (1911), p. 136. It is assumed here, by way of trial, that only emission from the atom takes place in quanta, but that absorption occurs continuously, an idea that later had to be given up. See especially Planck's *Scientific Autobiography* (New York: Philosophical Library, 1949) pp. 43–46.

LETTERS 9–18: EINSTEIN & SCHRÖDINGER

8. This sentence is written in the side margin of the letter, and was obviously added after its completion.

9. An exchange of letters on statistical mechanics had preceded these letters.

10. A. Einstein, *Berliner Berichte* (1925) p. 3.

11. A. Einstein, *Naturwissenschaften* 14 (1926), p. 300.

12. See A. Einstein, *Naturwissenschaften* 14 (1926), p. 222. [English translation in A. Einstein, *The World As I See It* (New York: Philosophical Library, 1935), p. 204.

13. This concerns the easily established fact that the tea leaves scattered at the bottom of a cup collect in the middle when the tea is stirred.

14. In a letter to N. Bohr dated 13 May 1928 Schrödinger thanks him for an offprint of Bohr's paper, *The Quantum Postulate and the Recent Development of Atomic Physics* (*Naturwissenschaften*, 16, [1928], p. 245). He remarks on the fact that the Heisenberg uncertainty relation prevents one from being able to distinguish between neighboring quantum states under certain circumstances, and mentions as examples the conjugate action and angle variables as well as the motion of a molecule in an ideal gas. He sees this as a limitation on the applicability of the *old* experimental concepts which will have to be replaced by a *new* system of ideas, which will, of course, be very difficult.

Bohr replies to this in a letter dated 25 May 1928, saying that he sees no basis for giving up the old concepts, and that all difficulties can be removed by means of the principle of complementarity. He does not consider the remark on angle variables to be sound because, in interpreting experiments with the help of the concept of stationary states, one always has to deal with those properties of an atomic system that depends upon the phase relations over *a large number* of consecutive periods. He does not quite understand the application of the uncertainty relation to a gas molecule because here the momentum quantity conjugate to the coordinate has no unique value.

15. See footnote 14 in the previous letter.

16. See Bohr's account of his discussions with Einstein in Bohr's book *Atomic Physics and Human Knowledge* (New York: John Wiley and Sons, Inc., 1958), p. 32.

17. Point Peconic, Long Island.

18. Schrödinger's witty conceptual experiment with a "smeared-out cat" is described in his article on the state of the quantum theory at that time in *Naturwissenschaften* 23 (1935), on page 812.

19. For Max Born's position on quantum mechanics, see his article, *The Interpretation of Quantum Mechanics* in his book *Physics in My Generation* (London and New York: Pergamon Press, 1956) p. 140.

LETTERS 19–21: LORENTZ & SCHRÖDINGER

20. Victor August Julius, born 1851, Professor of Theoretical Mechanics and Mathematical Physics at the University of Utrecht from 1896 on, died 1902.

21. The time dependence is that of a sinusoidal vibration; since, however, it could not be established in what form this factor was written down in the original the gap was left open. Compare Letter 6.

22. This is clearly a slip. Schrödinger evidently means (1) and the unnumbered equation $\ddot{u} = -\frac{4\pi^2}{h^2}E^2u$.

23. W. Heisenberg, *Mathematische Annalen* 95 (1926), p. 683.

24. P.A.M. Dirac, *Proc. Royal Soc.* A 109 (1925) p. 649; A 110 (1926) p. 561.

25. G. Wentzel, *Zeitschrift fur Physik* 37 (1926), p. 80.

26. The "festival day" referred to is the golden anniversary of Lorentz's doctorate, celebrated in Leyden in December 1925.

27. Schrödinger participated in the 4th Solvay Conference in Brussels in April 1924; Lorentz presided over these Conferences.

28. "Electron" probably appears here erroneously instead of "atom."

29. E. Schrödinger, *Naturwissenschaften* 14 (1926) p. 664.

30. Lorentz uses the symbol H for $h/2\pi$, now usually denoted by $\hbar$.

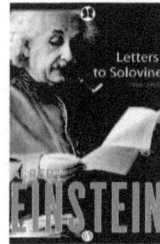

PHILOSOPHICAL LIBRARY

*Philosophical Library's mission is to reintroduce readers to **books of lasting value** by the intellectual icons of the twentieth century, including Albert Einstein, Jean-Paul Sartre, Kahlil Gibran, and André Gide.*

FIND OUT MORE AT

WWW.PHILOSOPHICALLIBRARY.COM

FOLLOW US:

@PhilLibrary

Facebook.com/PhilosophicalLibrary

PhilosophicalLibrary.Tumblr.com

OPEN ROAD

INTEGRATED MEDIA

Open Road Integrated Media is a digital publisher and multimedia content company. Open Road creates connections between authors and their audiences by marketing its ebooks through a new proprietary online platform, which uses premium video content and social media.

Videos, Archival Documents, and New Releases

Sign up for the Open Road Media newsletter and get news delivered straight to your inbox.

Sign up now at
www.openroadmedia.com/newsletters

FIND OUT MORE AT
WWW.OPENROADMEDIA.COM

FOLLOW US:
@openroadmedia and
Facebook.com/OpenRoadMedia